U.S. Department of Transportation
Federal Transit Administration

REMOTE INFRARED AUDIBLE SIGNAGE PILOT PROGRAM EVALUATION REPORT

September 2009

Prepared by:
Margaret Petrella, Lydia Rainville and David Spiller

The Volpe National Transportation Systems Center
Economic and Industry Analysis Division
55 Broadway
Cambridge, MA 02142

Report Number: **FTA-MA-26-7117-2009.01**

Sponsored by:
Federal Transit Administration
Office of Research, Demonstration and Innovation
U.S. Department of Transportation
1200 New Jersey Avenue, SE
Washington, DC 20590

Remote Infrared Audible Signage Pilot Program – Evaluation Report

REPORT DOCUMENTATION PAGE	*Form Approved* *OMB No. 0704-0188*

1. AGENCY USE ONLY (Leave blank)	2. REPORT DATE September 2009	3. REPORT TYPE AND DATES COVERED Final Evaluation Report (October 2007- September 2009)

4. TITLE AND SUBTITLE Remote Infrared Audible Signage Pilot Program Evaluation Report	5. FUNDING/GRANT NUMBER DTFT60-07-N-0058
6. AUTHOR(S Petrella, Margaret; Rainville, Lydia; Spiller, David	
7. PERFORMING ORGANIZATION NAME(S) AND ADDRESS(ES) The Volpe National Transportation Systems Center (The Volpe Center) Research and Innovative Technology Administration 55 Broadway Cambridge, MA 02142	8. PERFORMING ORGANIZATION REPORT NUMBER FTA-MA-26-7117-2009.01
9. SPONSORING/MONITORING AGENCY NAME(S) AND ADDRESS(ES) Federal Transit Administration Office of Research, Demonstration and Innovation Website [http://www.fta.dot.gov/research] 200 New Jersey Avenue, SE Washington, DC 20590	10. SPONSORING/MONITORING AGENCY REPORT NUMBER

11. SUPPLEMNTARY NOTES
Available Online [http://www.fta.dot.gov/research]

12a. DISTRIBUTION/AVAILABILITY STATEMENT Available From: National Technical Information Service (NTIS), Springfield, VA 22161. Phone 703.605.6000, Fax 703.605.6900, Email [orders@ntis.gov]	12b. DISTRIBUTION CODE TRI-10

13. ABSTRACT (Maximum 200 words)
This report presents evaluation findings on the Remote Infrared Audible Signage (RIAS) Pilot Program in the Puget Sound Region of Washington. The installation, demonstration and evaluation of RIAS were required by a provision in the Safe, Accountable, Flexible, Efficient Transportation Equity Act: A Legacy for Users. RIAS is an orientation and mobility technology designed to eliminate barriers to accessibility for people who are visually impaired or are cognitively or developmentally disabled. The evaluation was designed specifically to better understand the impacts of RIAS on multimodal accessibility, on transit ridership, on transit operators, and on quality of life issues. This report outlines the evaluation methodology and presents focus group and survey findings, lessons learned, conclusions, and recommendations.

14. SUBJECT TERMS Remote infrared audible signage, visual impairment, accessibility, Talking Signs	15. NUMBER OF PAGES 67
	16. PRICE CODE N/A

17. SECURITY CLASSIFICATION OF REPORT – Unclassified	18.SECURITY CLASSIFICTION OF THIS PAGE Unclassified	19. SECURITY CLASSIFICATION OF ABSTRACT Unclassified	20. LIMITATION OF ABSTRACT

NSN 7540-01-280-5500   Prescribed by ANSI Std. 239-18298-102   Standard Form 298 (Rev. 2-89)

Foreword

The evaluation of Remote Infrared Audible Signage (RIAS) was funded by the Federal Transit Administration and is authorized under a provision of the Safe, Accountable, Flexible, Efficient Transportation Equity Act: A Legacy for Users (SAFETEA-LU, Public Law 109-59). As per the legislative requirement, the U.S. DOT Volpe Center prepared a Report to Congress on its findings. That Report to Congress formed the basis for this Evaluation Report, with the addition of a set of lessons learned and a set of programmatic recommendations.

Disclaimer

This report is being disseminated under the sponsorship of the U.S. Department of Transportation, Federal Transit Administration, in order to foster information exchange.

The U.S. Government assumes no liability or responsibility for the contents of the report or the use of the report. The U.S. Government is not endorsing any manufacturers, products, or services cited herein and any trade name that may appear in the report has been included only because it is essential to the contents of the report.

Acknowledgements

The Volpe Center would like to acknowledge the guidance and assistance provided by Mr. Raj Wagley, Program Manager and Mr. Walter Kulyk, Director, FTA Office of Research, Demonstration and Innovation, and Mike Baltes, Director, FTA Office of Technology. In addition, the Volpe Center would like to thank all the members of the RIAS project team. In particular, Mr. Michael Miller and Ms. Ella Campbell, of Sound Transit, provided invaluable support with regards to the recruitment and training of participants and the administration of the screener questionnaires. Other team members, including Mr. Scott Kearney, of Hidalgo & DeVries, and Mr. W. David Marlin and Mr. Mark Talbot, of Lamoreaux McLendon, played a key role, and their efforts are greatly appreciated.

Finally, the Volpe Center would like to acknowledge the individuals who participated in the study; their comments and feedback were critical to the evaluation.

REMOTE INFRARED AUDIBLE SIGNAGE PILOT PROGRAM EVALUATION REPORT

CONTENTS

Executive Summary 1

1.0 Introduction 4
    1.1 The Technology 6
    1.2 The Installation 8
    1.3 Project Coordination 10

2.0 Methodology 11
    2.1 Institutional Review Board Approval Process 11
    2.2 Recruitment 12
    2.3 Extended Use Study 13
    2.4 Transit Operator Interviews 15
    2.5 Limitations of the Evaluation 15

3.0 Findings 17
    3.1 Background 17
    3.2 Effect on Multimodal Accessibility 19
    3.3 Effect on Operators and Their Passengers 23
    3.4 Effect on Ridership 24
    3.5 Effect on Education, Community Integration, Work Life and General Quality of Life 25

4.0 Conclusions 27
5.0 Recommendations 29
6.0 Lessons Learned 32

Appendix A: Project Timeline 35
Appendix B: RIAS System Components and Specifications 37
Appendix C: Number of Transmitters at Each Station 41
Appendix D: Study Materials 43

List of Figures and Tables

Figure 1: RIAS Transmitter 6
Figure 2: RIAS Receiver 6
Figure 3: Infrared Light Beam Emitted by the Transmitter 6
Figure 4: Woman Using RIAS at the Powell Street Station in San Francisco 7
Figure 5: Map of RIAS Installation 8
Figure 6: RIAS Controller Unit 9
Figure 7: ST Express Bus at Issaquah Park-and-Ride 18
Figure 8: Bus Bays and Sounder Platform at Everett Station 19
Figure 9: Bus Bay with RIAS Transmitter..20
Figure 10: View of Seattle Skyline over King Street Station 23
Figure 11: Sounder Commuter Rail by Puget Sound 25

Table 1: Transit Agencies Serving Stations with RIAS Deployment 8

Photo Credits

Cover Page Photos and Figures 6, 8-10: With permission of W.D. Marlin, Lamoreaux McLendon Telecommunications
Figure 1, 2: With permission of Talking Signs®
Figure 4: Courtesy of Dr. William Crandall
Figures 7, 11: Courtesy of Sound Transit

Executive Summary

The installation, demonstration, and evaluation of Remote Infrared Audible Signage (RIAS) was required by a provision in the Safe, Accountable, Flexible, Efficient Transportation Equity Act: A Legacy for Users (SAFETEA-LU, Public Law 109-59) that called for a RIAS pilot program. RIAS technology is a remote infrared communication system designed to eliminate barriers to accessibility for people who are blind, visually impaired, or cognitively or developmentally disabled by assisting them with both orientation and mobility (O&M). The RIAS system consists of permanently installed transmitters that emit signals by directional infrared light beams and handheld receivers that decode the signals into an audio message. By scanning the environment with the receiver, the user receives audible messages that label key features or provide directional information.

As outlined in the legislation, the evaluation of RIAS must include the following assessments:

1. The effect of the pilot program on multimodal accessibility in public transportation;
2. The effect of the program on operators of public transportation and their passengers;
3. The effect of making public transportation accessible to people with visual, cognitive, and learning disabilities on ridership of public transportation and the use of paratransit; and
4. The effect of the program on the education, community integration, work life, and general quality of life of the targeted populations.

Based on a competitive Request for Proposals (RFP) issued on July 1, 2006, the Federal Transit Administration (FTA) selected Central Puget Sound Regional Transportation Authority (Sound Transit) of Seattle, Washington to accomplish the installation and demonstration of RIAS. FTA signed a cooperative agreement with Sound Transit on September 6, 2006. This installation of RIAS is the first multimodal application in this country. Under the cooperative agreement, Sound Transit installed the RIAS technology (including 303 transmitters) at six multimodal transit stations in the Puget Sound area. The Federal Government provided $1.895 million, and Sound Transit contributed a 17% local share for a total cost of $2.283 million to install the RIAS technology at the six transit stations.[1] Originally, RIAS technology was to be installed at the Seattle—Tacoma International Airport, Overlake Transit Center, Lynwood Transit Center, and Auburn Station, as well as on buses; however budget constraints did not allow for installation at these stations or on buses.

---

[1] This total cost includes all equipment and installation costs.

Sound Transit issued an RFP for the installation work and awarded the contract on March 28, 2008 to Talking Signs Services, Inc. (TSSI). TSSI completed the installation of the system at the six transit stations in February 2009.

Under an agreement with the FTA, the United States Department of Transportation (U.S. DOT), John A. Volpe National Transportation Systems Center (The Volpe Center) conducted the evaluation of the RIAS pilot program in Seattle. The evaluation was designed specifically to address the requirements laid out in SAFETEA-LU. As per the legislative requirement, the U.S. DOT Volpe Center submitted a Report to Congress on its findings. That Report to Congress formed the basis for this Evaluation Report, with the addition of a set of lessons learned and a set of programmatic recommendations. The scope of the evaluation did not require a cost-benefit analysis, nor did it require a comparative analysis of RIAS technology with other technologies designed to provide wayfinding information for blind or visually impaired populations.[2]

For the evaluation, the Volpe Center recruited study participants (also referred to as "subjects" or "respondents" in this Report) who are blind or visually impaired to use RIAS during the normal course of their travels on public transportation for an extended period of time. Surveys, a focus group, and qualitative interviews were used to measure the impacts of RIAS among the recruited study participants. Overall, fourteen blind or visually impaired persons participated in the study.[3]

Based on findings from the evaluation, RIAS technology enhanced multimodal accessibility for nearly all study participants. In general, RIAS technology enabled users to navigate the transit system more efficiently and with greater confidence and independence, though users felt that the coverage and placement of transmitters was better at some stations than at others. Respondents indicated that the most important information provided by RIAS was the location of the bus bays and train platforms, with accompanying information on the bus numbers, directions, and destination. The installation of signs at bathrooms and drinking fountains was useful, but was deemed less of a priority.

While respondents had generally positive feedback on the effects of RIAS on multimodal accessibility, the deployment of the technology at only six transit stations and the lack of transmitters on buses limited the value of the system to users. Respondents noted that the technology needs to be deployed at more stations. Also, while respondents were generally pleased with the ability of the technology to direct them to the appropriate bus bay, users were missing the final piece of information they needed to enable a seamless multimodal trip – namely, information on approaching buses. Confirming these findings,

---

[2] For more detailed information on costs of the RIAS system and for a review of existing wayfinding technologies, please see "Remote Infrared Audible Sign (RIAS) Full Implementation of Model Accessibility Program Report," prepared by Lamoreaux McLendon Telecommunications, June 2009.

[3] In addition, a human factors study measuring the usability of the technology in a multimodal setting was conducted by Hidalgo & DeVries among the same pool of study participants. The findings are presented in a separate report, "Remote Infrared Audible Signage Model Accessibility Project: Human Factors Testing and Evaluation."

only two respondents said they would purchase a receiver (at a cost of $290) given the current deployment level of six transit stations, but nearly all respondents (except one) said they would purchase the receiver if the system were deployed at **all** transit stations and bus stops. Respondents also cited a number of improvements to the technology that would increase the value of the system, including a hands-free capability, Bluetooth compatibility, the use of GPS, and the incorporation of RIAS in cell phones.

This study found that participants' use of RIAS had no measurable effect on transit operators or their passengers. A number of factors, including the short evaluation period, the small sample of users, and the nature of the installation (i.e., no availability on buses), made it unlikely that transit operators or passengers would be impacted by the technology. Nonetheless, some respondents indicated that using the technology would cut down on their number of requests for travel information, suggesting that as the system is deployed at more stations or as the pool of users increases, this might decrease the workload for transit operators (e.g., they would receive fewer requests for information).

Due to the short evaluation period, the findings on the impact of RIAS on ridership are largely based on respondents' projections of how the technology would affect their travel behavior, rather than on actual changes in behavior. The findings on this question were mixed, with some respondents indicating that RIAS would increase their trip making behavior, and others saying it would have no effect. While less confident travelers were more likely to say that RIAS would impact their travel behavior, even confident travelers noted that they would be more likely to travel to unfamiliar cities if they knew RIAS were installed at airports and transit stations.

A few respondents indicated that RIAS had a positive impact on their quality of life during the evaluation period, through enhancing their independence and their confidence when using public transportation. Nearly all respondents, however, agreed that the deployment of RIAS would have to be more widespread in order for the technology to really have an impact on their life.

1.0 Introduction

The installation, demonstration, and evaluation of Remote Infrared Audible Signage (RIAS) was required by a provision in the Safe, Accountable, Flexible, Efficient Transportation Equity Act: A Legacy for Users (SAFETEA-LU, Public Law 109-59) that called for a RIAS pilot program. The key objective of the demonstration was to assess the benefits of remote infrared audible signage technology for people who are visually impaired, or cognitively or learning disabled. Congress directed the Secretary to submit a Report on the RIAS pilot program no later than September 30, 2009.

SAFETEA-LU Section 3046

(a) (6) Pilot programs for remote infrared audible signs.—

(A) In general.—For each of fiscal years 2006 through 2009, not less than $500,000 shall be made available by the Secretary to carry out a pilot program to determine the benefits of remote infrared audible signage technology for provision of wayfinding and information to people who are visually, cognitively, or learning disabled.

The requirements of the Report, as outlined in the legislation, included the following assessments:

1. The effect of the pilot program on multimodal accessibility in public transportation;
2. The effect of the program on operators of public transportation and their passengers;
3. The effect of making public transportation accessible to people with visual, cognitive, and learning disabilities on ridership of public transportation and the use of paratransit; and
4. The effect of the program on the education, community integration, work life, and general quality of life of the targeted populations.

The Federal Transit Administration (FTA) issued a Request for Proposals (RFP) on July 1, 2006 and received six proposals. An evaluation team at FTA reviewed the proposals and made the recommendation to award the cooperative agreement to the Central Puget Sound Regional Transportation Authority (Sound Transit) of Seattle, Washington. On September 6, 2006, FTA signed the cooperative agreement with Sound Transit to accomplish the installation and demonstration of RIAS (see Appendix A for a timeline of key project milestones). The Federal Government provided $1.895 million, and Sound Transit contributed a 17% local share for a total cost of $2.283 million to install the RIAS technology at the six transit stations. The average approximate cost of installation (including equipment cost) per transit station was $405,000.[4] Originally, RIAS

---

[4] King Street Station was omitted from this calculation, since most of the transmitters at King Street Station were installed in 2003 as part of a demonstration project (only 14 transmitters were installed as part of the current project).

technology was to be installed at the Seattle-Tacoma International Airport, Overlake Transit Center, Lynwood Transit Center, and Auburn Station, as well as on buses; however budget constraints did not allow for installation at these stations or on buses.

Sound Transit issued an RFP for the installation work and awarded the contract on March 28, 2008 to Talking Signs Services, Inc. (TSSI), who completed the installation of the system at the six transit stations in February 2009.

In support of the FTA, the United States Department of Transportation (U.S. DOT), John A. Volpe National Transportation Systems Center (The Volpe Center) conducted the evaluation of the RIAS pilot program. The evaluation was designed specifically to address the requirements of the legislation; the scope of the evaluation did not include a cost-benefit analysis, nor did it include a comparative analysis of the RIAS technology with other technologies designed to provide wayfinding information for blind or visually impaired populations.

1.1 The Technology

RIAS is a remote infrared communication system designed to eliminate barriers to accessibility for people who are blind, visually impaired, or cognitively or developmentally disabled by assisting them with both orientation and mobility. The technology, developed at California Pacific Medical Center's Smith Kettlewell's Eye Research Institute, utilizes spoken infrared message transmissions to provide wayfinding information. For people who cannot read signs or have difficulty doing so, RIAS provides the same type of wayfinding information, but in an audible format.

Figure 1: RIAS Transmitter

Figure 2: RIAS Receiver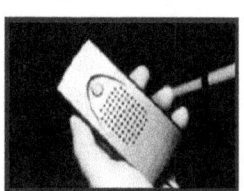

RIAS consists of two key components:

1) permanently installed transmitters that emit audio signals by directional infrared light beams (Figure1); and 2) handheld receivers (Figure 2) that decode those signals into an audio message.

The beam starts at the infrared diode located within the transmitter head and spreads out in a cone shape, becoming wider as it moves away from the source (see Figure 3). Through adjusting the transmitters, including the light emitting diode (LED) arrays, it is possible to control the maximum distance at which the message is received, the direction in which the message is transmitted, and the area that the message covers.[5]

Figure 3: Infrared Light Beam Emitted by the Transmitter

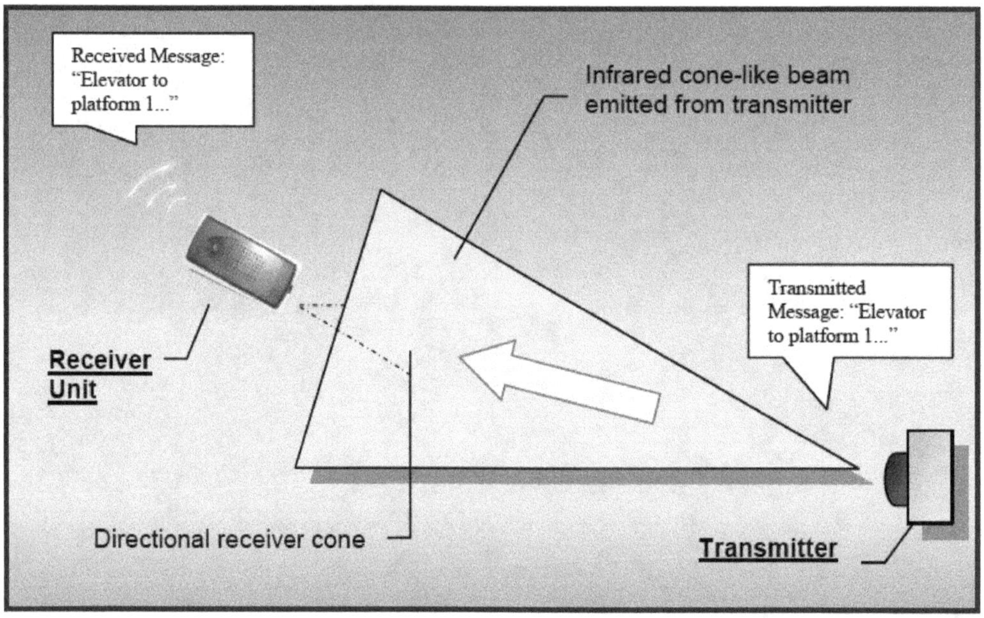

---

[5] See Crandall, William, Billie Louise Bentzen, Linda Meyers, and John Brabyn. "New Orientation and Accessibility Option for Persons with Visual Impairment: Transportation Applications for Remote Infrared Audible Signage," *Clinical and Experimental Optometry*, 84.3, 2001.

To use the system, the user scans the environment with a handheld receiver, depressing a small button located on the receiver. As the receiver picks up the signal from a transmitter, that signal is decoded and translated into an audible message. The message is delivered through a speaker in the receiver, or can be accessed through a headset. Users pick up the digitally prerecorded message when the front of the receiver is pointed in the direction of the infrared transmitter, and the messages become clearer and more intense as the user approaches the transmitter (e.g., the sign).

The transmitters are installed so that they label key features in the environment, such as bathrooms, telephones, elevators, stairs, ticket vending machines, and bus or train platforms. In addition, transmitters are placed so that they can provide users with directional information (e.g., "To Bus Bay 6"). (See Appendix B for more detailed information on the technology.)

RIAS has been deployed in a number of U.S. cities, at select transportation centers and buildings, as well in selected transit stations including the Bay Area Rapid Transit (Powell Street Station and Fremont Station), San Francisco Municipal Railroad (selected stops), and Capital Area Transit Authority in Lansing, Michigan (on all buses). These agencies operate RIAS in support of a single mode of transportation (bus or rail). The demonstration of RIAS in the Puget Sound area, however, is the first multimodal application that seeks to provide a seamless connection among different modes of transportation.

Figure 4: Woman Using RIAS at the Powell Street Station in San Francisco

1.2 The Installation

Under its cooperative agreement with FTA, Sound Transit installed 303 transmitters at six transit stations in the Puget Sound area (see Appendix C for the number of transmitters installed at each station).[6] The six transit stations (shown right) are:

- Everett Station
- King Street Station[7]
- Federal Way Transit Center
- Kent Station
- Tacoma Dome Station
- Bellevue Transit Center

Figure 5: Map of RIAS Installation

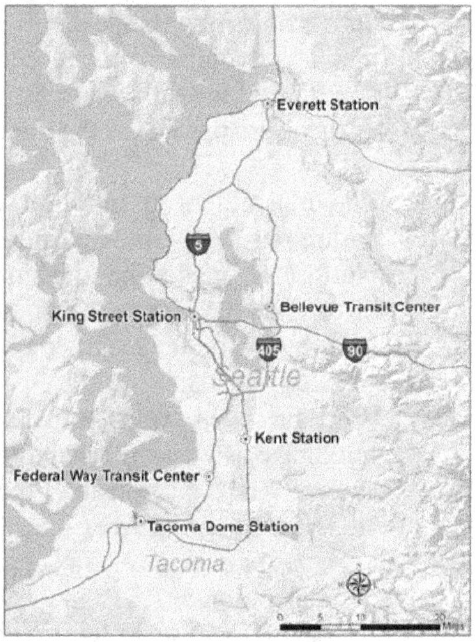

Sound Transit operates approximately 23 transit facilities (excluding Park and Rides). Sound Transit selected these six stations for the RIAS installation based on one or more of the following considerations: the station is multimodal, has a high level of services including transfers between systems, has transfer options from paratransit to fixed route, or is among the busiest facilities in the system. As Table 1 illustrates, most of the stations have both bus and commuter rail service ("Sounder"), and three of the stations also provide Amtrak service.

Table 1: Transit Agencies Serving Stations with RIAS Deployment

Bellevue Transit Center	Everett Station	Federal Way Transit Center	Kent Station	King Street Station	Tacoma Dome Station
Sound Transit (express bus)	Sound Transit (express bus)	Sound Transit (express bus)	Sound Transit (express bus)	Sounder Commuter Rail	Sound Transit (express bus)
King County Metro Transit (bus)	Community Transit (bus)	King County Metro Transit (bus)	King County Metro Transit (bus)	Amtrak	Pierce Transit (bus)
	Everett Transit (bus)			Sound Transit (express bus)	Sounder Commuter Rail
	Island Transit (bus)	Pierce Transit (bus)	Sounder Commuter Rail		
	Skagit Transit (bus)				Amtrak
	Sounder Commuter Rail				Greyhound
	Amtrak				
	Greyhound				

---

[6] Crosswalk installations are planned for Everett Station and King Street Station (eight transmitters total), but were not installed in time for the evaluation. Twenty-seven transmitters will also be installed at the Amtrak platform at King Street Station.

[7] In 2003 Sound Transit conducted a demonstration project of RIAS at King Street Station and twenty-four of the transmitters installed for that project were functional during the current RIAS pilot program.

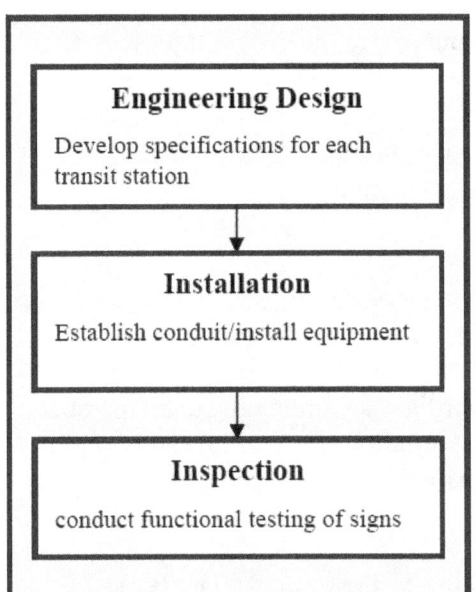

Engineering Design
Develop specifications for each transit station

Installation
Establish conduit/install equipment

Inspection
conduct functional testing of signs

Sound Transit, in coordination with its subcontractors, developed the engineering design specifications for each transit station, including the placement of the transmitters, source of power or low voltage sources, the content of the messages, and the areas where the signs would be pointed. As part of this process, Sound Transit considered noise levels at each of the stations and safety issues, as well as structural barriers that might affect the transmission of the infrared signals. A certified O&M specialist made site visits to all six stations and provided input to the design specifications.

The installation of the RIAS technology required establishing conduit as well as the cabling for power/data to the transmitters and mounting hardware for the transmitters. The conduit for the transmitters was either existing or new, as specified in the engineering design. The transmitters are manufactured in two different types. One type, the stand-alone, has the message built in and only requires 120 VAC power cabling for operation. The second type, known as the control type, uses low voltage data cabling for power and for transmitting the programmed message to the transmitter. Each controller unit can house up to 12 programmed messages, and multiple controller units are located within a central cabinet (see Figure 6). In sum, this central cabinet houses multiple messages which connect to multiple control-type transmitters.

During installation, all signs were inspected for coverage areas and messages, and a final inspection was conducted to insure the working status of all signs.

Figure 6: RIAS Controller Unit

1.3 Project Coordination

The RIAS project required a great deal of coordination among the different project team members, including:
- Sound Transit
- Lamoreaux McLendon (Sound Transit's project management team)
- Hidalgo and DeVries (FTA's project management team)
- The Volpe Center evaluation team

Bi-weekly team meetings, organized by Hidalgo and DeVries, enabled team members to provide status updates and to raise issues and concerns regarding the project.

Two key factors contributed to the need for close coordination among the team members. First, project team members shared responsibility for certain project tasks, and second, schedule constraints necessitated ongoing communication. Each of these factors is described in further detail below.

Shared project responsibilities: Sound Transit and the Volpe Center each played a role in the recruitment of study participants and thus had to closely coordinate their efforts. At the outset of the project, Sound Transit was assigned primary responsibility for recruitment, due to the fact that its staff had well established contacts within the targeted study population. In addition, since Sound Transit was on-site, its staff could more easily recruit potential participants by visiting local organizations or attending chapter meetings. The Volpe Center provided guidance on recruitment materials and assisted Sound Transit by contacting relevant organizations.

In the same fashion, coordination was necessary to implement the human factors study. The Volpe Center provided the overall design of the human factors study, while Hidalgo and DeVries (in cooperation with Lamoreaux McLendon and Sound Transit) conducted the study and wrote the report.

Schedule Constraints: As outlined in SAFETEA-LU, FTA was required to submit A Report to Congress on the RIAS demonstration project no later than September 30, 2009. To meet its deadline, FTA needed the Volpe Center Report by April 30, 2009. At the same time, the Volpe Center could not begin its evaluation until the installation of the technology was completed. Given these schedule constraints, it was of paramount importance that the project team be kept apprised of progress with the installation, particularly changes to the schedule. In addition, it was critical to be updated on any changes to the scope of the installation, as these might affect the evaluation.

2.0 Methodology

The evaluation of RIAS was specifically designed to address the four research questions outlined in the legislation, namely the effect of the pilot program on: multimodal accessibility in public transportation on; operators of public transportation and their passengers; ridership of public transportation and the use of paratransit; and the education, community integration, work life, and general quality of life of the targeted populations. To measure these effects, the research team determined that subjects should use RIAS during the course of their normal travels on public transportation for an extended period (a minimum of eight weeks). In order to obtain reliable measures, subjects would have to incorporate the technology into their "real-world" travel routines. To this end, an "extended use" study was developed, whereby subjects were given a receiver to use independently when traveling on public transportation. Surveys, as well as a focus group, were utilized in the extended use study to better understand the impacts of RIAS on multimodal accessibility, on ridership, and on quality of life issues. To measure the effects of RIAS on transit operators and their passengers, the research design called for qualitative interviews with transit operators.

A human factors study measuring the usability of the technology in a multimodal setting was conducted by Hidalgo & DeVries. The findings are presented in the report, "Remote Infrared Audible Signage Model Accessibility Project: Human Factors Testing and Evaluation." For the human factors trial, subjects participated in a one-day test held at transit stations where RIAS is installed. The human factors trial and the extended use study used the same pool of participants. On the day of the human factors trial, subjects were trained on the technology, and as part of the trial, researchers asked participants to perform specific tasks using the technology and collected data on their performance.[8] At the end of the trial, participants who qualified for the extended use study were given a receiver for their personal use.

The following section of the Report describes the Institutional review Board (IRB) approval process, recruitment methods, and the methodology for the extended use study and the transit operator interviews. Limitations of the evaluation are also described.

2.1 Institutional Review Board Approval Process

Studies involving the use of human subjects generally require the approval of an accredited Institutional Review Board (IRB). The purpose of the IRB is to ensure that the rights and the welfare of the study participants are being protected throughout the research project. For the RIAS project, the Volpe Center utilized the Washington State Institutional Review Board (WSIRB) at the Washington State Department of Social and

---

[8] Due to scheduling issues, the training schedule was staggered, with trainings occurring between February 7 and March 6, 2009.

Health Services. Due to the minimal risk of the RIAS project, the Volpe Center was able to utilize the expedited approval process. Under expedited review, only two WSIRB members review and approve the application package (in contrast to convening a meeting of the full Review Board).

To obtain approval, the Volpe Center completed an extensive application outlining the research purpose, methodology, and procedures, as well as potential risks to the subjects and steps taken to mitigate those risks. In addition to the application, the Volpe Center had to submit all materials being used in the study, including study advertisements, recruitment screeners, consent forms, surveys, discussion guides, and scripts for any type of contact with respondents. As part of the approval process, Volpe staff also had to participate in human subjects protections training, offered via an online course that took approximately 6 to 8 hours to complete.

In August 2008, the Volpe Center submitted its initial application package for both the human factors study and the extended use study. The WSIRB staff responded with significant comments and revisions. Based on feedback from the WSIRB, the Volpe Center, in consultation with FTA and Hidalgo and DeVries, decided to submit two separate applications; the Volpe Center prepared the application package for the extended use study, and Hidalgo and DeVries prepared the application package for the human factors study. Through an iterative process working with the WSIRB, the Volpe Center refined the application and the study materials to meet the requirements of the WSIRB. Final approval was obtained in December 2008. Unfortunately, during the WSIRB review process (August through December), recruitment of study participants was prohibited, which likely contributed to the lower than anticipated number of respondents.

2.2 Recruitment

Sound Transit and the U.S. DOT Volpe Center worked with local agencies and advocacy organizations for the blind to develop a subject pool for the study. These organizations included Lighthouse for the Blind, Washington State Department of Services for the Blind, local chapters of the National Federation of the Blind (NFB), the American Council of the Blind (ACB), and King County Developmental Disabilities Division. The U.S. DOT Volpe Center and Sound Transit distributed a recruitment notice describing the study to these organizations, as well as to orientation and mobility specialists working in the Puget Sound area, and requested that they distribute the notice to their clientele and "spread the word" about the study. In addition, Sound Transit posted information about the study on its website, and the Sound Transit Accessibility Manager attended several chapter meetings of the NFB and the American Council of the Blind and made a presentation at the Lighthouse for the Blind to describe the study and to solicit participants. All interested participants were asked to complete an eligibility form.

In order to participate, potential recruits had to meet the following requirements:

> - Subjects must be blind, visually impaired, cognitively impaired, or developmentally disabled
> - Subjects must be 18 years of age or older
> - Subjects cannot have a hearing disability
> - Subjects must use public transportation independently (i.e., able to travel without a human guide)[9]
> - Subjects must use at least one of the six transit stations where RIAS is deployed[10]

The Volpe Center, in coordination with Hildago and DeVries, designed a recruitment screener that was used to confirm participants' eligibility (See Appendix D). Sound Transit staff administered the screener to each potential respondent by telephone. Recruitment efforts yielded an initial subject pool of 35 blind or visually impaired respondents. However, when telephone calls were made to determine eligibility, a significant number of potential respondents were either ineligible, unavailable to participate during the study period, no longer interested, or could not be reached (despite repeated attempts). In all, 14 respondents participated in the extended use study.

As part of the screening process, Sound Transit staff set up appointments with all eligible respondents interested in participating in the study. These appointments were used to obtain the respondents' informed consent to participate, as per the requirements of the WSIRB. Sound Transit staff also trained participants on how to use the technology and distributed RIAS receivers to respondents participating in the extended use study.

2.3 Extended Use Study

The original research design for the extended use study included the use of surveys and qualitative interviews. The goal was to administer at least 75 surveys (or more, pending recruitment) and to conduct qualitative interviews with a small subset of the respondents. This design was slightly altered in October 2008, when the Volpe Center learned that the Office of Management and Budget (OMB) Paperwork Reduction Act applied to the RIAS project.[11] According to the OMB Paperwork Reduction Act, surveys administered to ten or more persons must go through the OMB approval process and receive clearance before being administered. The approval process generally takes a minimum of six months. FTA and the Volpe Center considered two options:

---

[9] Ideally, the study would have included persons who were not currently using public transportation, but who were interested in becoming transit users. However, project funding was not sufficient to cover the training required for new transit users.

[10] This was a requirement of the extended use study, but not the human factors study.

[11] The Volpe Center thought that the RIAS project was exempt from the OMB Paperwork Reduction Act, in accordance with the exemption for the National ITS Program (see Section 5305(i)(2) of SAFETEA-LU); however, the ITS exemption did not apply to RIAS.

1) Proceed with the OMB approval process
    - Maintain current methodology
    - Seek an extension for the Report to Congress to accommodate the extra time required for the OMB approval process
    - Bolster recruitment efforts in tandem with the OMB survey review
2) Do not proceed with OMB approval
    - Revise the methodology so it does not violate the Paperwork Reduction Act
    - Maintain the current project timeline

In consultation with the U.S. DOT Volpe Center, the FTA decided for a number of reasons to pursue option 2. Based on the relatively small number of recruited participants at that point, it would be relatively easy to alter the methodology to conform to the Paperwork Reduction Act. Moreover, there was no guarantee that additional recruitment efforts would result in a significantly greater number of study participants. Finally, the FTA felt it was important to maintain the current schedule and deliver the Report to Congress by September 30, 2009, as mandated in the legislation.

With the approval of the FTA, the Volpe Center tailored its evaluation protocol to meet the requirements of the Paperwork Reduction Act. From the respondent pool of visually impaired and blind people, nine people were randomly selected for the survey pool. Those respondents not assigned to the survey pool were invited to participate in a facilitated discussion group. This methodology enabled the collection of both quantitative and qualitative data, with the focus group serving as a cross-check to the survey and enabling the research team to validate the survey findings. The focus group also offered the opportunity to hear the participants talk about their experiences with RIAS in greater detail.

For the survey effort, Volpe staff administered the surveys by telephone both before and after the deployment of RIAS. Six of the fourteen participants completed both a pre- and a post- survey. The pre-survey was used to gather contextual and background data on the respondents, including basic demographic information, current use of different transportation modes, level of confidence when using public transportation, and challenges faced in using public transportation. The post-survey asked respondents to assess RIAS. This included the impact of RIAS on multimodal accessibility, on overall trip making behavior, on level of confidence when traveling, and on general quality of life. The survey also probed respondents on the benefits of the technology, as well as problems or concerns.

The remaining eight subjects participated in a focus group discussion. The U.S. DOT Volpe Center staff developed a discussion guide that was used to assess the impacts of RIAS on multimodal accessibility, trip-making behavior, and quality of life issues. The focus group was convened on March 23, 2009 at Union Station, in Seattle. A U.S. DOT Volpe Center staff person moderated the discussion, and a court reporter was present to record verbatim responses from the participants. Of the eight respondents assigned to the focus group protocol, six were able to attend the focus group session. The U.S. DOT Volpe Center contacted the two participants who were unable to attend to obtain their feedback on the technology.

2.4   Transit Operator Interviews

To measure the impact of RIAS on transit operators and their passengers, the research design incorporated qualitative interviews with station agents. Since 2008, Sound Transit has assigned station agents to Sounder stations in an effort to enhance customer service. These station agents canvass the platforms, answering travelers' questions and assisting passengers as necessary. The U.S. DOT Volpe Center research team, in consultation with Sound Transit, determined that RIAS was more likely to have an effect on station agents than on transit operators, since station agents are "on the front lines" and can observe passengers using the technology. Moreover, because RIAS was *not* deployed on buses or trains, and therefore could not assist travelers in identifying specific bus numbers, the research team anticipated that transit operators were not likely to be impacted by the technology (e.g., even with RIAS, transit operators would continue to receive questions such as, "Is this bus number 532?"). As a result, the hypothesis developed for this research was that station agents should receive fewer requests for information or assistance from visually impaired travelers after the deployment of RIAS, compared to before deployment.

Evaluators conducted interviews with station agents at the end of the evaluation period after study participants had the opportunity to use the technology during the course of their normal travels. The U.S. DOT Volpe Center developed a qualitative interview guide to probe the number and types of requests that transit operators typically receive, how often they receive requests for information from blind or visually impaired customers, whether they had noticed any changes in the number or types of requests received in the last few months, their awareness of RIAS, and their assessment of the usefulness of RIAS.

2.5   Limitations of the Evaluation

Due to schedule delays with the awarding of contracts and the installation work, the RIAS installation was not completed until February 2009 (with the exception of crosswalks), which reduced the amount of time for the evaluation. Study participants only had four to six weeks to use the technology independently during the normal course of their travels. In addition, for some respondents, bad weather during the course of the evaluation period (including a snow storm) further limited their use of public transportation. Due to the short evaluation period of this study, it was difficult for the RIAS technology to have a measurable impact on multimodal access, on transit operators, or on the general quality of life of the study participants.

Another limitation of the study is that non-random recruitment methods were utilized, so it is not possible to generalize the findings from this study to the larger population of blind and visually impaired persons living in Seattle or in the nation. Project constraints necessitated a more passive, non-random approach to recruitment. Given the relatively small size of the population of interest for the study (blind or visually impaired persons who are independent users of at least one of the six transit stations where RIAS is deployed), conducting a telephone survey of all Seattle-area residents to identify potential recruits would have been cost-prohibitive. In addition, lists of Seattle-area residents who are blind, visually impaired or developmentally disabled do not exist (or are not publicly available), so it was not possible to randomly select subjects from such a list. Due to the small sample size, qualitative rather than quantitative analyses of the data were conducted.

Finally, despite targeted recruitment efforts, persons with cognitive or learning disabilities did not volunteer to participate. As a consequence, this study does not address the usefulness of RIAS to persons with cognitive, learning, or developmental disabilities. Interestingly, one focus group participant offered her opinion (unprompted) that RIAS would be useful beyond the blind community: *"This just doesn't help the blind population. It helps dyslexic people, and it helps…the elderly or the vets…It can help most people, not just blind folk."*

Additional research would be needed to assess the extent to which RIAS could benefit these subpopulations.

3.0 Findings

Section 3 begins with a presentation of descriptive background information on the survey respondents, based on their responses to the pre-survey. This background section is followed by separate sections for each of the four RIAS evaluation areas: effect on multimodal accessibility; effect on operators and their passengers; effect on ridership; and effect on education, community integration, work life and general quality of life.

3.1 Background

Demographic Characteristics

Of the fourteen participants who completed the extended use study, six were totally blind and eight were legally blind or partially sighted. For the six participants assigned to the survey, additional background information was collected. Of these six, only one was visually impaired from birth; the remaining five first became visually impaired between nine and twenty-nine years ago. The sample included a range of participants of whom four were female and two male. One participant was between the ages of 30 and 39, four were between 40 and 49, and one was between 50 and 64. The level of education completed varied; one participant had completed a high school degree, one had some college education, three had bachelor's degrees, and one had an advanced degree.

All six survey participants used technology to assist with both reading and mobility. The effect of the visual impairment on reading ability varied between the participants; two could read large print with the use of an aid and four could not read large print at all. All six had some ability to read Braille and used assistive technologies such as screen readers, closed-circuit televisions, and magnifiers for reading. For travel aids, all six used a long-white cane as a travel aid and one also used a Miniguide.[12] Before participating in the extended use study, four respondents had heard of RIAS, of whom three had used it. Only one participant described himself as "very familiar" with RIAS.

Use of Transportation

All six participants had been using transit independently for more than seven years, and all but two had received some kind of formal training (e.g., orientation and mobility, cane training). Overall, the participants were very confident using public transportation and traveling in familiar transit centers, slightly less confident making transfers at transit stations, and least confident in unfamiliar transit centers.

When asked about the most challenging aspects of using public transportation, all six responded that the most difficult aspects of travel were finding the right bus and bus stop.

---

[12] While all surveyed used long-white canes for mobility, it is worth noting that three of the focus group participants traveled with service dogs.

Other challenges included stop announcements, reading signs, riding Sounder, and figuring out when the next bus or train would arrive.

Figure 7: ST Express Bus at Issaquah Park-and-Ride

Overall, the six survey respondents were regular users of public transportation, though two were in periods of transition and noted that their answers did not reflect their typical use. In the month prior to the pre-survey, all six had taken metro buses, four had taken express buses, and three had each ridden the Sounder and dial-a-ride. Only two rode the Tacoma Link Light Rail (occasionally) and none used private van services. Four had taken taxis and all six had been in a private car. Two had regular access to private vehicles through their households. Three respondents frequently had access to private cars, and the others had access from two to six times a month. All rode the bus at least eight times a month and two rode the bus up to 25 times a month. The most frequent users rode public transit for trips to work or to school. Other frequent trips were for shopping and recreational activities.

Of the stations with RIAS installed, King Street was the most frequently used and was also used by the most respondents. The only other stations with a frequent user were Tacoma Dome and Bellevue—Everett and Federal Way were used occasionally, and no one used Kent. The stations where the most users transferred were King Street and Tacoma Dome. The lowest reported confidence levels were generally at King Street Station, but one participant also gave a low confidence rating at Everett and Tacoma Dome Stations.

Two participants agreed that difficulties related to public transportation limited their participation in social and recreational activities. All but one was at least somewhat satisfied with their ability to access public transportation and all but one responded that difficulties using public transportation made their lives more stressful.

Use of RIAS

During the evaluation period, all six survey respondents used the technology, with four using it only a few times. Of the remaining two, one used it a minimum of twice a week and the other used it about four times a week. Four of the eight focus group participants reported using the receiver two to three times a week, three used it a few times during the evaluation, and one did not use it outside of the initial training. On a scale of one to ten (with one indicating not at all satisfied and ten indicating extremely satisfied), survey respondents rated their experience with RIAS between a four and ten with an average rating of 6.5, though one commented that she would have given it an eight had it been in more places.

3.2 Effect on Multimodal Accessibility

The RIAS installation in Seattle was specifically designed to provide wayfinding information in a multimodal environment, so that users could more easily navigate the transportation network and obtain improved access to different modes of transportation. Multimodal transit stations can be particularly challenging to visually impaired persons, as they are often complex environments with multiple levels. The image below shows both the bus bays and the Sounder platform at Everett Station.

Figure 8: Bus Bays and Sounder Platform at Everett Station

Overall the study found that RIAS enhances accessibility, but with deployment at six stations only, its value is limited. The human factors trial, conducted by Hidalgo & DeVries, Inc., reported similar findings. In that study, most users were able to locate all of the signs for their assigned tasks, and they generally found the technology to be either very or somewhat effective and easy to use. However, a few respondents did experience difficulty using the system.

In the extended use study, both the survey and focus group findings show that RIAS enhances multimodal transfers by enabling travelers to locate transfer points more quickly—by increasing travel speed and allowing for faster orientation. One participant noted that it particularly enhanced transfers in unfamiliar stations. Another participant noted that *"if you're halfway familiar, you can reorient much more quickly."* A third participant, who has a short time to transfer when commuting, said: *"If the 500 isn't late, it's like a two or three minute window I have to get to the 574. Well, it's a lot nicer when I can whip out of the bus, and I know the general direction I need to go… and then I can just hold that little receiver up there and just tell Frieda go for it, and we can run along that platform and make good time to get there and make my transfer."*

RIAS increased the participants' confidence while making transfers as well. During the post-survey interviews, all but one respondent said that RIAS generally increased their confidence level when making transfers from one type of transportation to another. The survey findings showed, however, that its ability to enhance transfers varied from station to station. For example, one participant specified that the technology increased confidence at Bellevue and Tacoma Dome Stations, but was actually a hindrance and decreased his confidence at King Street Station, due to poor placement and inadequate coverage. During the human factors trial, participants also raised concerns about the limited coverage at King Street Station.

At Tacoma Dome, Bellevue, and Everett, more than half of survey participants who used those stations agreed that the technology was helpful or very helpful; the results were more mixed at King Street, Kent, and Federal Way. Survey findings also suggested that those who were less familiar with a station generally found RIAS to be more helpful at that station than those who were already familiar.

One participant also said RIAS would increase the number of transfers he would be willing to make. Currently, he uses a longer route for some destinations in order to minimize (or eliminate) transfers. With RIAS, however, he speculated that he would be more willing to take a direct route with transfers: *"I would tend to go and make more transfers where I knew the system was in place."*

Figure 9: Bus Bay with RIAS Transmitter

Both the survey and focus group confirmed that the types of information that were most valued, and which most enhanced multimodal accessibility, were signs indicating bus bays (pictured at left) or train platforms, with information on bus numbers, direction, and destination. In particular, participants stressed the importance of bus number information (e.g., which bus numbers stop at each bay). This type of detailed information was included for some bus bay signs but not all, and participants indicated the provision of such information should be consistent across all transit stations. Other information on the location of bathrooms, drinking fountains, and so on, while useful, was less of a priority. Supporting this, one focus group participant said: *"I mean the bathrooms and drinking fountains are wonderful, but if I had to pick and choose which signs I wanted…bus numbers and bay numbers would be more important."*

More broadly, RIAS helped with multimodal accessibility by facilitating all travel. More information makes travel easier, particularly, according to two survey respondents, at stations like Bellevue Transit Center which have many bus bays. One participant discovered a new route to navigate around King Street Station and another discovered the Greyhound bus station. Another participant also noted that RIAS would be helpful in situations where bus stops or bus routes have changed (assuming transmitters are updated with the new information in real-time).

RIAS also provided both survey and focus group participants with independent reassurance and confirmation as to where they were while traveling. As one noted, *"…even if someone tells me I'm standing at Gate B, that may not always be the case, where the receiver would absolutely confirm that."*

While the response to the effects of RIAS on multimodal accessibility was generally positive, users did find some limitations and drawbacks. One person felt she wasn't learning anything new from the receivers, so she stopped using it. Another indicated that after using the receiver several times, she felt she could navigate her way through the station without the receiver: *"and when I found I wasn't discovering anything really new, I just didn't – because it was hard with the dog and all the stuff I had to carry, and so like the last two weeks or so, I haven't done anything with it."* However, some still found the information useful due to a "convenience factor."

Other issues cited in both the survey and focus group included:

- <u>Lack of direction and distance provided in the signs.</u> Some users felt that the signs did not provide sufficient directional information (e.g., north or south, or in some cases, right or left), and that it would be helpful to have approximate distances (in feet) to the destination.

- <u>Inadequate placement of the transmitters.</u> One participant said that at some stations, particularly Bellevue and Tacoma Dome, it seemed like the transmitter placement was laid out by someone blind or visually impaired, but at other stations it did not. As another respondent indicated, *"There needs to be some care in how signs are placed."*

- <u>Limited coverage of stations and transit providers</u>. At Everett, only the Sound Transit stops had transmitters—the Community Transit and Everett Transit stops did not. Several participants also expressed concern that unmarked crossings (some diagonal) were not always labeled.

To assess the value of the system to users, evaluators presented participants with several different scenarios or trade-offs on cost versus coverage. Under the first scenario respondents were told the cost of the receiver ($290) and asked if they would purchase the receiver based on the current deployment of RIAS at the six transit stations. Only two survey participants said they would definitely buy it if they could afford it, and one focus group participant said she might. One who said yes cited the helpfulness of the information, particularly at night or in unfamiliar transit centers. Those who said they would not purchase the receiver cited limited information and features, and some said that they did not visit the stations frequently enough to justify the purchase. As a participant noted, "*I want to see some improvements in it before I pay that kind of money, in the receiver itself.*"

According to users, an enhancement that would increase the value of the system (and perhaps result in the purchase of a receiver under the first scenario) includes a hands-free capability, so that users do not have to depress a button to receive the transmitted message. A number of participants mentioned that carrying the receiver was a burden, since one hand is already occupied with a cane or a service dog, and they may be carrying other items as well. Another participant said that having to scan the environment with the receiver made him feel conspicuous: "*I'd rather not look kind of silly if I don't have to.*" He said he would prefer to wear the receiver around the neck (for example, on a lanyard). Participants also suggested placing the receiver on a visor or on the end of a cane to enable "hands-free" use. Other ideas for enhancing RIAS include Bluetooth compatibility, the use of global positioning system (GPS) technology, and the incorporation of RIAS in cell phones.

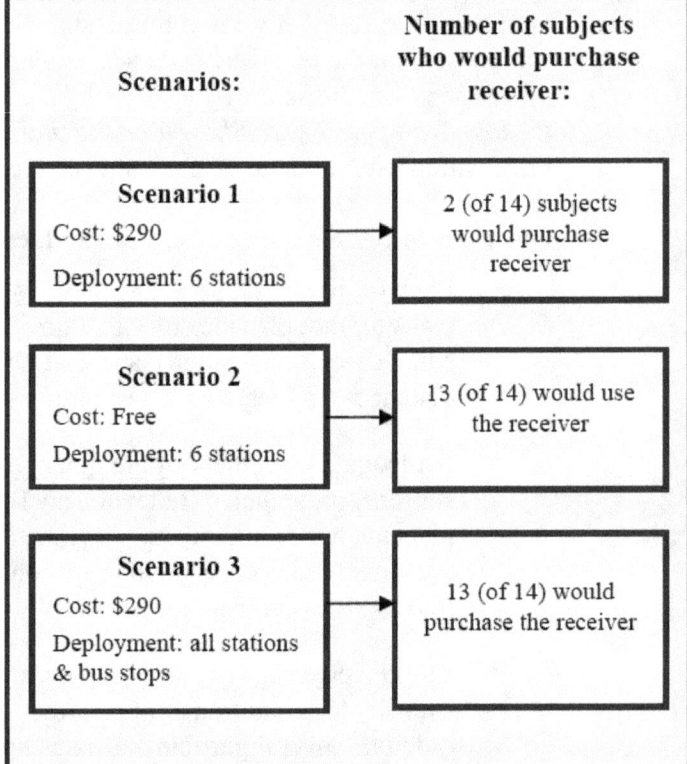

Under the second scenario, participants were told that RIAS deployment would still be at the current six stations, but that receivers would be free. Under these circumstances, all persons surveyed and all but one focus group participant said they would use it, though two would be willing to give up receivers to those who might need them more or travel more frequently in the six stations. Another said RIAS would have to be greatly improved at stations like King Street and Kent.

For the third scenario, participants were told that the receiver would cost $290, and the system would be deployed at all transit stations and transit stops in the Puget Sound area.[13] Under this scenario, all six surveyed said they would purchase it, mainly citing independence and confidence, enhanced travel, and the usefulness of bus stop information as primary factors in the decision. Similarly, all but one of the focus group participants indicated they would likely purchase the receiver if the system were deployed throughout the Puget Sound region. As

---

[13] There are approximately 40-50 transit stations in the Puget Sound area and on the order of 10,000 bus stops throughout the three-county region.

one focus participant claimed, "*We need more of them, more signs. I would definitely pay the money if I knew it was all over the Puget Sound.*" The one participant who would not purchase the receiver said he would rather hire someone or bring a friend, but would consider it if it were part of a cell phone.

Figure 10: View of Seattle Skyline over King Street Station

Participants overwhelmingly agreed that RIAS would be more useful and more likely to enhance multimodal accessibility if deployed system-wide at transit stations and bus stops in Seattle. Participants also suggested equipping buses with transmitters so users could identify each bus. As one focus group participant explained, "*…If we could just point at a bus…that would be a lot more helpful because it's awesome that we know where buses stop so we're at the right bus stop. But then the problem is getting on to the right bus.*"

In addition, some participants would like RIAS to be available and interoperable in cities across the U.S., both in transit stations and airports. As one focus group participant noted, "*I would buy it in a heartbeat…if I knew it was at the airports throughout the country.*"

3.3 Effect on Operators and Their Passengers

Deployment of RIAS was recently operational and the extended use study was only underway for six weeks with a limited number of respondents. Thus, the evaluation was unlikely to yield significant findings on the effects of RIAS on transit operators or passengers. A more extended period of evaluation would be required for potential users to learn about the technology, and to use it on a regular basis, in order to measure effects on transit operators and their passengers.

Evaluators conducted short, qualitative interviews with station agents at King Street Station, Tacoma Dome Station, Everett Station and Kent Station. Based on these interviews, RIAS did not have an effect on station agents. The station agents were aware of the deployment but had only not seen the technology being tested by contractors installing the system, not being used. Station agents speculated that the technology would be useful, but rarely saw visually impaired travelers during their shifts.

Evaluators also tried to address this question indirectly, by asking respondents whether RIAS would reduce the frequency with which they asked others for information, including transit operators or other passengers; however, because of the limitations of the study, some of the answers provided were based more on projection than actual experience. Two survey participants said that using RIAS had and would have no effect on whether they would ask transit operators or station agents for information. Three survey respondents said that they would be less likely to ask for information and four were less likely to ask for information from other passengers. One noted that with RIAS he would know basic information and so would probably be less likely to ask other passengers. However, as one participant pointed out, because RIAS does not provide information on which bus is at a stop, *"once you're at the bus stop, you would still have to ask the driver what bus it is."* Both survey and focus group participants made this observation.

Providing additional insight into this question, several survey respondents and focus group participants cited independence as the primary benefit of RIAS; in particular, they would not have to depend on other people to be present, be willing to help, or provide reliable information.

Overall, the survey and focus group participants indicated that using RIAS would cut down on their number of requests for information. *"I would ask a lot less questions…I would never have to ask for the right bays. If you labeled, like I say, if I'm in the tunnel stations here, if you labeled the stairs down to northbound platform, southbound platform, I'd never have to ask that."*

3.4 Effect on Ridership

Participants were asked whether their use of RIAS had any impact on their trip-making behavior, such as the number or types of trips made. Because of the limited time frame of the evaluation, some of the trips made during the evaluation period represent a technology effect. That is, the surveys and focus group revealed that in some cases, respondents were visiting stations with RIAS simply because they were curious and wanted to test the technology. Due to the limited time frame of the study, it is unclear whether RIAS would have any long-term impacts on travel behavior. As a result, the answers may reflect projections of future use as well the limited experience during the extended use study period. Limitations in coverage also made interpreting the results of these questions more difficult, as some of the responses may reflect the limited coverage of the technology rather than limitations of the technology itself. Several respondents noted that the stations they use most frequently did not have RIAS installed.

Overall, the study found mixed results regarding the effect of RIAS on ridership. Five survey respondents and three focus group participants said that RIAS would increase their trip-making behavior. This finding is supported by anecdotal evidence from the surveys—one respondent used RIAS to ride Sounder, the commuter rail, for the first time, and another made a trip that he/she doesn't make frequently to visit a friend at one of the RIAS stations, saying that the trip is normally difficult because it involves many connections.

Figure 11: Sounder Commuter Rail by Puget Sound

For both the survey and focus group, the findings seemed to show that for more confident travelers, RIAS generally would not affect trip-making behavior. One noted that *"it wouldn't change how I plan a trip or change where I go at all. It would just enhance where I go and how I get there maybe."* Less confident travelers were more likely to say that RIAS would affect their trip-making behavior, though one confident traveler reported, *"if this was available nationwide or what not, I would definitely travel a lot more."*

3.5 Effect on Education, Community Integration, Work Life and General Quality of Life

The limited time frame for the evaluation made it difficult to measure the impacts of the RIAS on quality of life issues. When survey and focus group respondents were asked if the technology had an impact on their quality of their life, most said that deployment would have to be more widespread for it to have an impact. *"But if you're just going to do it a few select places, no, it's not going to affect my life."* Another responded, *"We need more of it if it's going to be worth its salt. For me, to get where the RIAS signs exist is like a 30 mile adventure."* The same respondent continued, *"In the long haul, the only way it's going to improve my life is if there's more of it."* Reflecting this, some of the answers may represent projections of future use of RIAS in addition to the limited experience gained during the extended use study.

Of the six survey respondents, four believed RIAS did or could have an impact on their social or recreational opportunities. Reasons given were that RIAS provided an expanded travel area and an additional tool to improve travel skills—examples respondents provided included easier shopping and an ability to make social trips that require a number of transfers. One of the respondents who did not believe that the deployment had an impact said it would have a significant effect if improved. In the post-survey conducted for the human factors trial, one participant felt that RIAS would be very useful for social or recreational trips which require travel to unfamiliar places, and another participant noted that, *"For education it is important to be on time, so [RIAS] would be helpful."*

Five of the survey respondents believed RIAS did or could have an impact on their quality of life. Four of them cited independence as the primary benefit. Other benefits included confidence, security, and convenience. One summarized the benefits as increased information on new places and more information on familiar stations. One participant in the focus group said that it *did* affect the quality of her life, because she used the receivers at two stations she had not previously visited (Tacoma Dome and Bellevue), and as a consequence, she now feels comfortable going to those transit stations. But at the same time, she indicated that in order for her to use RIAS more, the system would have to be deployed at more stations. Another participant noted, *"There's so much potential there with airports and shopping malls and tunnels, and the list goes on and on and on if it's all compatible."*

4.0 Conclusions

A provision in SAFETEA-LU, Public Law 109-59, required the installation, demonstration, and evaluation of RIAS. Specifically, the legislation called for an assessment of four key questions:

- The effect of the pilot program on multimodal accessibility in public transportation;
- The effect of the program on operators of public transportation and their passengers;
- The effect of making public transportation accessible to people with visual, cognitive and learning disabilities on ridership of public transportation and the use of paratransit; and
- The effect of the program on the education, community integration, work life, and general quality of life of the targeted populations.

Based on findings from the study, RIAS enhanced multimodal accessibility for nearly all participants. In general, RIAS enabled users to navigate the network more efficiently and with greater confidence and independence, though users felt that the coverage and placement of transmitters was better at some stations than at others. Respondents indicated that the most important information provided by RIAS was the location of the bus bays and train platforms, with accompanying information on the bus numbers, directions, and destination. The installation of signs at bathrooms and drinking fountains was useful, but was deemed less of a priority. Respondents also felt RIAS would be particularly useful in providing real-time updates on bus stop or bus route changes.

While respondents had generally positive feedback on the effects of RIAS on multimodal accessibility, the deployment of the technology at only six transit stations and the lack of transmitters on buses limited the value of the system to users. Respondents noted that the technology needs to be deployed at more stations. Also, while respondents were generally pleased with the ability of the technology to direct them to the appropriate bus bay, users were missing the final piece of information they needed to enable a seamless multimodal trip–namely, information on approaching buses. Confirming these findings, only two respondents said they would purchase a receiver (despite noting cost concerns) given the current deployment level of six transit stations, but nearly all respondents (except one) said that they would purchase the receiver if the system were deployed at all transit stations and bus stops. Respondents also cited a number of improvements to the technology that would increase the value of the system, including a hands-free capability, Bluetooth compatibility, the use of GPS, and the incorporation of RIAS in cell phones.

With regard to the effect of RIAS on transit operators or their passengers, this study did not detect an impact. A number of factors, including the short evaluation period, the small sample of users, and the nature of the installation (i.e., no availability on buses), made it unlikely that transit operators or passengers would be impacted by the technology. Nonetheless, some respondents indicated that using the technology would cut down on their number of requests for information, suggesting that as the system is deployed at more stations or as the pool of users increases, this might decrease the workload for transit operators (e.g., they would receive fewer requests for information).

Due to the limitations of the evaluation, the findings on the impact of RIAS on ridership are largely based on respondents' projections of how the technology would affect their travel behavior. The findings on this question were mixed, with some respondents indicating that RIAS would increase their trip making behavior, and others saying it would have no effect. While less confident travelers were more likely to say that RIAS would impact their travel behavior, even confident travelers noted that they would be more likely to travel to unfamiliar cities if they knew RIAS were installed at airports and transit stations.

A few respondents indicated that RIAS had a positive impact on their quality of life during the evaluation period, through enhancing their independence and their confidence when using public transportation. Nearly all respondents, however, agreed that the deployment of RIAS would have to be more widespread in order for the technology to really have an impact on their life. As one respondent noted, *"In the long haul, the only way it's going to improve my life is if there's more of it."*

5.0 Recommendations

Prior to moving forward with a significant investment in the RIAS technology, further evaluation is needed to determine the usefulness and the market viability of the technology (particularly given the significant costs of deploying RIAS system-wide). Conducting a technology scan, assessing the use and performance of RIAS in Seattle over the next year, as well as gathering data on how RIAS has performed in other cities, would provide FTA with valuable information on which to base future investment decisions.

The following section of the report outlines the U.S. DOT Volpe Center's programmatic recommendations in more detail.

- **Conduct a technology scan and convene a panel of experts.** Before investing in RIAS, FTA should consider performing an in-depth review of available O&M technologies. In support of FTA, the Volpe Center is currently conducting a Transit Wayfinding Technologies Assessment for all transit passengers, including passengers with disabilties. The Wayfinding Technology Assessment will provide the field operational test framework for the demonstration deployment of state-of-the art-technologies at 2 to 3 metropolitan areas. FTA should consider building on this related study by obtaining more detailed information on technologies designed to assist persons who are blind or visually impaired.

    The proposed technology scan should address the following questions:

    - What is the range of available technologies designed to assist blind/visually-impaired populations with wayfinding information? What are the cutting edge technologies?
    - What are the relative strengths and weaknesses of each technology?
    - Where have these technologies been deployed or tested? If evaluations have been conducted at these sites, what are their findings?
    - What are the costs of each technology?

    As part of the technology scan, the research team should determine what progress has been made in the development of the RIAS system over the last five years, and what improvements in the technology are on the horizon (e.g., will they be developing a hands-free receiver soon)? In addition, it is important to determine what evaluation studies, if any, have been conducted regarding the deployment of RIAS on buses or trains. This assessment should be designed to provide FTA with information on how RIAS stands relative to other O&M technologies, both in terms of its capabilities and cost, and if possible, its performance.

    To complement this work, Volpe recommends that FTA (perhaps in conjunction with the Transportation Research Board (TRB), American Public Transportation Association (APTA), and/or the Access Board) convene a panel of experts to obtain guidance on the "state of the art" regarding wayfinding technologies for persons who are blind or visually impaired. The panel of experts could also assist with recommendations of which technologies should be field demonstrated as part of the Wayfinding Technology Assessment, ensuring that the technologies are useful for blind or visually impaired people.

Based on the data collected in the technology scan and the recommendations of the expert panel, FTA will receive important guidance on the most promising wayfinding technologies for blind or visually impaired people.

- **Increase awareness of the RIAS system in the Puget Sound area.** In order for RIAS to be used, the populations who might benefit from the technology (blind, visually impaired, cognitively disabled) need to be aware that the system has been deployed in their area. FTA should work with Sound Transit, Washington State Department of Services of the Blind, and with advocacy organizations in Seattle (Lighthouse for the Blind, local chapters of the ACB and NFB) to increase awareness of RIAS. Potential ideas include:

    o Conduct RIAS demonstrations at chapter meetings of ACB or NFB

    o Convene RIAS training sessions, coordinated through the Washington State Department of Services for the Blind and/or Lighthouse for the Blind

    o Lease RIAS receivers on a monthly basis (suggested by a study participant)

- **Continue to monitor use and performance of RIAS in Seattle**. While this report provides some valuable insights on RIAS, the findings are limited due to the short evaluation period and the small number of respondents. FTA should continue to evaluate the use and performance of RIAS over the next year (assuming efforts are made to increase awareness of the availability of the technology in the Puget Sound area). Suggested areas for data collection include:

    o Track sales of receivers in Puget Sound area (and leasing of receivers, if a leasing strategy is used)

    o Track customer feedback on the technology (ask Sound Transit to keep a log of any feedback)

    o Conduct qualitative interviews with relevant personnel at Lighthouse for the Blind and Washington State Department of Services for the Blind, as well as other advocacy organizations (and potentially O&M specialists) to obtain feedback on RIAS. If substantial efforts have been made to promote RIAS, and the response from the targeted populations is lukewarm, then these interviews could provide insight on why adoption of RIAS has not been more widespread.

- Track performance of the technology (e.g., How often is maintenance of the signs performed? How often are signs not working?)

- **Collect available data on RIAS use and performance in other cities.** In addition to tracking the use and performance of RIAS in Seattle, determine what data (if any) is available for other cities where RIAS has been installed (in the U.S. or abroad). It would be useful if FTA could obtain data on receiver sales in other cities, or evaluation data that is based on longer term use of RIAS (vs. data from human factors testing).

- **Be aware of current efforts to update accessibility guidelines.** The Access Board is updating its accessibility guidelines for buses and vans covered by the Americans with Disabilities Act (ADA). The proposed revisions would require new vehicles in excess of 22 feet in length operated by medium and large public entities on fixed-route systems be equipped with automated interior stop and exterior route announcement systems.

  As the RIAS study demonstrated, blind and visually impaired travelers place a premium on obtaining information on approaching buses (information not provided by the current deployment of RIAS in Seattle), so these guidelines will certainly provide valuable wayfinding information. However, the automated announcement may have limited usefulness if the announcement is only made once when the doors of the bus are opened. The automated announcements must be repeated (at some designated interval), so that passengers who are not within earshot of the first announcement can still catch the second or third announcement. Additional research among blind or visually-impaired passengers is needed to clarify the impacts and potential benefits of the new accessibility guidelines. Similarly, it would be useful to understand the benefits of deploying RIAS on buses or trains.

- **Consider conducting a study among developmentally or cognitively disabled populations.** Unfortunately, the current evaluation was not able to recruit persons with developmental or cognitive disabilities. In order to determine the benefits of RIAS for this population, a separate study would need to be conducted. Based on experience from the current evaluation, recruitment for such a study will be time consuming and will require significant resources (including multiple site visits to make presentations).

6.0 Lessons Learned

The following section of the report outlines two sets of key lessons learned from the evaluation. The first set of lessons learned pertains to the evaluation process and is intended to provide guidance to evaluators across a broad range of evaluation studies. The second set of lessons learned addresses the installation and maintenance of RIAS and is specific to that technology.

Evaluation Process

- **Document the role of each of the partners.** The RIAS project was a collaborative effort, with Sound Transit and its subcontractors providing assistance with certain aspects of the evaluation. In such cases, it is important to clearly define and document the roles and activities of each partner, and to revise the document, if necessary, during the course of the evaluation. This will ensure that there is no confusion about who is doing what (and why).

- **Ensure adequate time and funding for the OMB clearance and the IRB approval processes.** Surveys administered to ten or more individuals must receive clearance from the OMB. This process includes the completion of an application package and takes a minimum of 6 months (and in most cases even longer). Similarly, studies involving human subjects must receive approval from an accredited IRB. The extensive IRB research application required several weeks to complete. Moreover, the IRB approval process was an iterative one, taking approximately 4 months from initial submission to the receipt of final approval. In addition to building in adequate time and funding, studies requiring OMB clearance and/or IRB approval should initiate these processes as early as possible in their schedules, since recruitment of subjects cannot begin until the research has been approved.

- **Devote significant resources to recruitment efforts.** For the RIAS project, recruitment was a collaborative effort, with Sound Transit assuming the main responsibility and the Volpe Center providing assistance. Although Sound Transit posted information about the evaluation on its website and attended meetings of local advocacy organizations for the blind and visually impaired, the recruitment numbers were low.[14] Sound Transit did not have sufficient time or staffing to recruit respondents more aggressively. The Volpe Center also made numerous calls to local advocacy organizations, but did not have sufficient funds to assume a larger role with recruitment. Multiple site visits by the evaluation staff, for example, likely would have bolstered recruitment.

    In future studies, evaluators must ensure adequate funding and time for recruitment. This is particularly important when conducting studies of specific subpopulations that may be hard to reach or whose contact information is not available. In such cases (e.g. RIAS), establishing relationships with local

---

[14] Recruitment rates were also affected by the IRB approval process, as all recruitment efforts had to be halted until the research was approved by the WSIRB.

advocacy organizations or groups is critical and is an extremely time-consuming process.

- **Use multiple methods of recruitment**. Multiple methods of recruitment are important to bolstering recruitment rates and to obtaining a diverse set of respondents. As previously mentioned, Sound Transit used both online tools (its website) and in-person methods (attending local advocacy group meetings) to recruit potential respondents. In addition, Sound Transit had planned to intercept blind people at transit stations, but unfortunately, due to the lack of time and resources, this recruitment strategy was not pursued.

- **Use multiple evaluation methods.** The evaluation team used both surveys and focus groups to collect feedback from study participants, and this proved to be an effective methodology. The focus groups complimented the survey data by providing a more in-depth perspective on some of the issues and concerns raised in the survey. By collecting both types of data, the evaluation team had a richer understanding of participants' use of and attitudes toward RIAS technology.

- **Ensure a sufficiently long evaluation period.** Due to time constraints associated with the submission of the Report to Congress, participants only had 4 to 6 weeks for the evaluation (e.g., to use the technology independently during their travels). While this timeframe would be sufficient to measure overall opinions about the technology, including effectiveness and usability, it was inadequate for addressing the evaluation question at hand, namely the effect of RIAS on quality of life, work and education. When collecting feedback from users, it became evident that a number of respondents were using the technology out of curiosity and had not yet incorporated use of RIAS into their normal travel routines. To measure changes in quality of life, users need to use the technology for a much longer period of time (on the order of 4 to 6 months).

- **Market the technology before initiating the evaluation.** Following installation or deployment of a technology, there should be a period of at least 2-4 months before the evaluation begins. During this period, marketing of the technology can occur, so that potential users can learn about the technology. These marketing efforts can also serve as an important recruitment tool for evaluation of the technology. In the case of the RIAS project, Sound Transit ensured that local advocacy organizations were aware of the deployment, but it did not have the resources to pursue a systematic marketing of the technology (e.g. brochures, pamphlets, etc.).

- **Plan for delays with contracting and installation.** With the RIAS project, there were significant delays in contracting the services of TSSI, the firm that installed the technology. Sound Transit issued its RFP in March 2006 and the contract was awarded one year later (see Appendix A). Similarly, there were delays with the installation. The installation, originally slated for completion in September 2008, was not completed until February 2009. With the deployment of new technologies, evaluation teams must assume that delays are likely and must be able to adapt their evaluation schedule and methodology to accommodate such delays.

Installation/Maintenance of RIAS Technology

- **Utilize a team of blind people to assist with the placement of transmitters at station.** An O &M specialist walked through each of the transit stations and provided Sound Transit with critical assistance in determining the placement of the transmitters. In addition, as some respondents indicated in the evaluation, it would be useful to have a small team of blind and visually-impaired persons provide similar assistance.

- **Consult with advocacy organizations on which transit stations and bus stops should be equipped with RIAS.** Local advocacy organizations may be able to provide useful data about the transportation patterns of their clientele, including which transit stations or bus stops are more likely to serve blind or visually impaired populations. If RIAS is more widely deployed in Seattle, or deployed in other U.S. cities, this data (if it exists) should be used to assist in the selection of stations and/or bus stops to be equipped with RIAS.

- **Perform frequent maintenance checks of the transmitters.** During the evaluation period, there were one or two RIAS transmitters that were not functioning. Sound Transit should periodically check that all transmitters are functioning according to specification.

Appendix A: Project Timeline

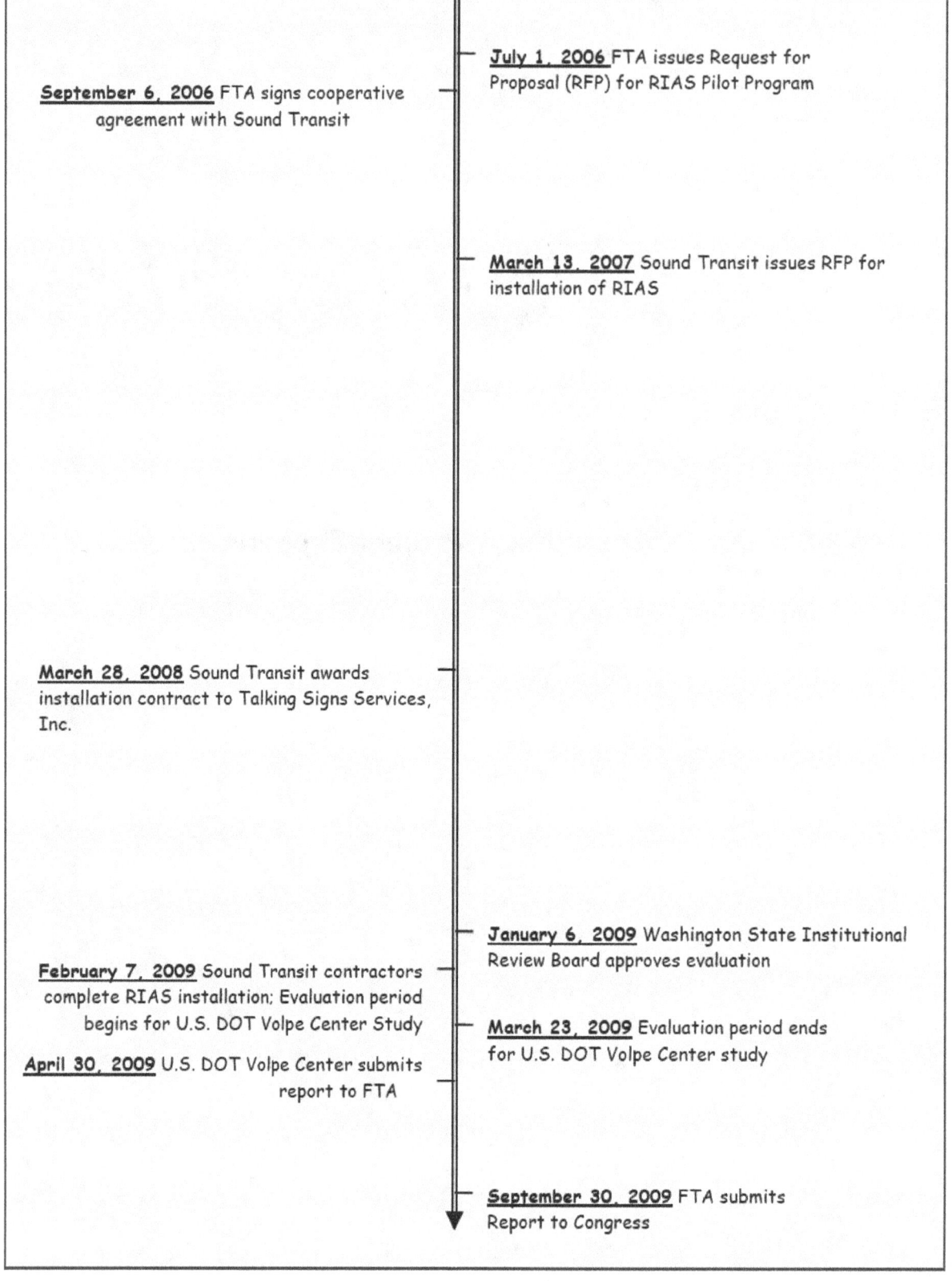

Appendix B: RIAS System Components and Specifications

Remote Infrared Sign Components

1. **Talking Sign LVC Controller Unit**
    - Houses up to 12 Type 3B Transmitter Cards for producing vocal messages using infrared technology
    - Programmable message lengths of up to 30 seconds
    - Provides 4 output drivers for a total of 24 LED diodes

2. **Talking Sign 324 Series Transmitter Head**
    - Transmits recorded vocal messages via infrared technology

3. **Talking Sign 406 Type 3A Stand Alone Transmitter: Specifications**

Specification	Description
Range	Maximum 65.6 feet
Angle	Up 45 degrees/Down 45 degrees Left 45 degrees/Right 45 degrees
Output	Infrared Beam/provides 4 output drivers for a total of 24 LED diodes
Voice	Self recording available Maximum 30 seconds
Power Input	100 volt /115 volt AC, 0.1 A
Dimensions	4.7 inches W x 4.7 inches D x 2.8 inches H
Weight	17.6 ounces

4. Talking Sign 407 Type 3A Stand Alone Transmitter: Specifications

Specification	Description
Range	Maximum 65.6 feet
Angle	Up 45 degrees/Down 45 degrees Left 45 degrees/Right 45 degrees
Output	Infrared Beam/provides 4 output drivers for a total of 24 LED diodes
Voice	Self recording available Maximum 30 seconds
Power Input	230 volt AC, 0.1 A
Dimensions	4.9 inches W x 4.9 inches Dx3.2 inches H
Weight	2.2 pounds
Attachment	Fuse, 250 volt

5. Talking Sign Type 2 Receiver: Specifications

Specification	Description
Receiver Sensitivity	About 65.6 feet
Infrared Receiving Angle	About 40 degrees
Speaker	Circle type, diameter of 2 inches, 8 ohm
Output Terminal	Ear Phone Jack (Mono Mini-Jack)
Power Supply	9 volt dc (battery 006P type 1 unit)
Operating Time	With speaker: 4 hours With earphones: 12 hours
Dimensions	2 inches W x ¾ inch D x 4.5 inches H
Weight	5 ounces
Attachments	Strap for hanging from neck/ear phone

Appendix C: Number of Transmitters at Each Station

Station	Number of transmitters
Bellevue Transit Center	21
Everett Station	97
Federal Way Transit Center	37
Kent Station	57
King Street Station	14
Tacoma Dome Station	77

Appendix D: Study Materials

Pre-Survey
Post Survey
Focus Group Discussion Guide

RIAS EXTENDED USER EVALUATION
PRE- SURVEY FOR BLIND AND VISUALLY IMPAIRED SUBJECTS

Hello, may I please speak with < first name / last name>? (*MUST speak to the participant*)

A. IF PARTICIPANT IS NOT AVAILABLE, IDENTIFY YOURSELF AND REQUEST A CALLBACK TIME:

Initial Call #	Date/Time	Callback Request #	Date/Time
Call 1	/	Callback 1	/
Call 2	/	Callback 2	/
Call 3	/	Callback 3	/
Call 4	/	Callback 4	/

B. IF PARTICPANT NEEDS TO COME TO THE PHONE:
Hi, am I speaking with < first name / last name>? (*Proceed with Introduction*)

IF PARTICIPANT ANSWERED PHONE:
INTRODUCTION
Hi, < first name>, my name is **(your first and last name)**. I'm calling on behalf of the U.S. Department of Transportation to ask you some questions about your use of transportation in the Seattle area. This is part of the study on Remote Infrared Audible Signage. Your participation in this interview is completely voluntary. Also, please remember that your responses are confidential, and your personal information, such as your name, will not appear in any of the reporting of the data.

Would this be a good time to ask you some questions?

[IF <YES>: Great. (Continue with interview below)]

[IF <NO, This is not a good time>: Can we schedule a more convenient time in the next few days?] <Record alternate date and time>_____

[IF <NO, I do not wish to participate>: Thank you for your time. We will no longer call you about this study. (Terminate call)]

I'd like to start off by gathering some background information.

Q. 1   Which of the following best describes your visual disability?

☐ Partially-sighted or low vision, or
☐ Legally-blind, or
☐ Totally blind

Q. 2   At what age did you become visually impaired?

_____ years old

☐ Check here if visually impaired since birth

Q. 3   Which one of the following best describes your ability to read:

☐ I can read large print without any aids
☐ I can read large print with an aid
☐ I cannot read large print at all

Q. 4   Can you read Braille?

☐ Yes
☐ No

Q. 5   Do you use any technologies to assist you in reading?

☐ Yes → Please specify:_____
☐ No

Now I'd like to ask you some questions about transportation and mobility

Q. 6   Which, if any, of the following travel aids do you use? [READ] As I read each one, let me know.
☐ Long or white cane,
☐ Guide Dog or other service animal,
☐ Electronic Travel aid (e.g. laser cane); IF YES: Please specify:_____
☐ Scooter
☐ Other (please specify:_____) OR
☐ None

Q. 7 Before learning about this study, had you heard of Remote Infrared Audible Signage, also known as Talking Signs?

☐ Yes -- GO TO Q. 8
☐ No -- GO TO Q. 9

Q. 8 Have you ever used Remote Infrared Audible Signage?

☐ Yes – GO TO Q. 8a
☐ No – GO TO Q. 9

Q. 8a How experienced are you with using Remote Infrared Audible Signage? [READ CATEGORIES]

☐ Very experienced
☐ Somewhat experienced
☐ Not very experienced
☐ Not at all experienced

Q. 9 How long have you been using transit independently? [READ CATEGORIES]

☐ Less than one year
☐ 1-3 years
☐ 4-6 years
☐ 7 or more years

Q. 10a What type of travel training have you had?
_____
_____

Q. 10b How long was your travel training_____

Q. 11 Using a seven point scale, where "1" means **not at all confident** and "7" means **very confident**, please rate your level of confidence in each of the following areas [emphasize 7 point scale]:

	Not at all Confident						Very Confident
a. Using public transportation	1	2	3	4	5	6	7
b. Making transfers from one type of transportation to another type of transportation	1	2	3	4	5	6	7
c. Traveling in unfamiliar transit centers	1	2	3	4	5	6	7
d. Traveling in familiar transit centers	1	2	3	4	5	6	7

Q. 12 What are some of the most challenging or difficult aspects of using public transportation for you? [PROBE: Anything else?]

_____
_____
_____
_____

Q. 13 A. Which of the following types of transportation do you use in the Seattle area? [START WITH ITEM a. IF USED, ASK B. CONTINUE WITH ITEMS b-j, AND FOR EACH ITEM THAT IS USED, FOLLLOW IMMEDIATELY WITH B]

B. (FOR EACH TYPE USED) Last month, approximately how many days did you use [INSERT ITEM FROM Q. 13 A]

	A		B
	Use	Do not Use	# Days last month
a. Public transit bus (e.g. King County Metro)	1	0	_____
b. Express bus (e.g. Sound Transit)	1	0	_____
c. Amtrak	1	0	_____
d. Sounder Commuter Rail	1	0	_____
e.. Dial-a Ride or door-to-door paratransit service	1	0	_____
f. private van services (e.g. Lighthouse for the Blind)	1	0	_____
g. Tacoma Link Light Rail	1	0	_____
h. Taxi	1	0	_____
i. Private car (family or friend)	1	0	_____

→ **IF USE:** Do you have access to a private vehicle through a household member or a friend? _____

j. Other? Please specify:

_____	1	0	_____
_____	1	0	_____

Q.14 A. For each of the following types of trips, please indicate whether or not you ever use public transportation to make that trip. [START WITH ITEM a. IF TRIP TYPE IS MADE, ASK B. FOR EACH ITEM (a-g), ASK A AND B TOGETHER]

B. (FOR EACH TRIP TYPE CHECKED IN 'A':) In the past month, how many times did you use public transportation to make [INSERT TRIP TYPE CHECKED IN 'A']. Please count a round trip as a single trip.

	A Ever Use:		B # trips on public
	Yes	No	trans. last month
a. Commute to/from work	1	0	_____
b. Other work related trips	1	0	_____
c. Travel to/from school	1	0	_____
d. Shopping	1	0	_____
e. Appointments, such as medical, personal	1	0	_____
f. Recreation or social	1	0	_____
g. Are there other trip types for which you use public transportation (specify below)			
_____	1	0	_____
_____	1	0	_____

Q. 15 Which of the following transit stations do you currently use? [READ ENTIRE LIST – RECORD BELOW UNDER Q. 15]
[FOR EACH TRANSIT STATION USED IN Q. 15, ASK Q. 16, Q. 17, and Q 18 TOGETHER]

Q. 16. How many days in a typical month do you use [INSERT TRANSIT STATION USED IN Q. 15]? [RECORD BELOW UNDER Q. 16].

Q. 17 Do you switch from one type of transportation to another type of transportation at [INSERT TRANSIT STATION USED IN Q. 15]? For example, do you switch from Amtrak to the bus, from a car to the bus, from one bus to another bus, from paratransit to the bus etc.… [RECORD BELOW UNDER Q. 17. IF YES: Could you describe the transfer you make at this station? RECORD UNDER "DETAILS"]

Q. 18 Using a seven point scale, where "1" means **not at all confident** and "7" means **very confident**, please rate your level of confidence in finding your way through [INSERT TRANSIT STATION USED IN Q. 15. REPEAT FOR EACH TRANSIT STATION USED AND RECORD BELOW UNDER Q. 18]

		Q.15 Use	Q.15 Do Not Use	Q. 16 # Days Used Typical Month	Q. 17 Transfer	Q. 17 No Transfer	Q.18 Confidence Rating (1-7)
a.	King Street Station	1	0	_____	1	0	_____
	Q. 17 Details on transfers:						
b.	Everett Station	1	0	_____	1	0	_____
	Details on transfers:						
c.	Bellevue Transit Center	1	0	_____	1	0	_____
	Details on transfers:						
d.	Kent Station	1	0	_____	1	0	_____
	Details on transfers:						
e.	Federal Way Transit Ctr	1	0	_____	1	0	_____
	Details on transfers:						
f.	Tacoma Dome Station	1	0	_____	1	0	_____
	Details on transfers:						

Q. 19 In the past month, have you made any unfamiliar trips using public transportation?

☐ Yes → In the past month, about how many unfamiliar trips did you make? _____
☐ No

Q. 20 In a typical month, how many times do you ask a **transit operator or a station agent** for travel-related assistance or information? For example, information on which bus you are boarding, schedule information, directions to another bus or train, etc. [Please estimate the total number of requests during a typical month] _____

Q. 21 In a typical month, how many times do you ask **other transit passengers** for travel-related assistance or information? For example, information on which bus you are boarding, schedule information, directions to another bus or train, etc.? [Please estimate the total number of requests during a typical month]_____

Q. 22 Do you ever avoid trips or activities because of difficulties related to the use of public transportation?

☐ Yes → In an average month, how many trips would you say you avoid? _____

☐ No

Q. 23 Using a seven point scale where "1" means completely disagree and "7" means completely agree, please indicate the extent to which you agree or disagree with the following statements:

		Completely Disagree						Completely Agree
a.	Difficulties using public transportation limit my participation in social and recreational activities	1	2	3	4	5	6	7
b.	I am satisfied with the access I have to different types of transportation in the Seattle area, including train, bus etc.	1	2	3	4	5	6	7
c.	Difficulties using public transportation make my life more stressful	1	2	3	4	5	6	7

Now I just have a few final questions.

Q. 24 What is the highest level of education you have received? [DO NOT READ]

☐ Grade School
☐ Some high school
☐ High school graduate
☐ Vocational school
☐ Some college
☐ Associate's degree
☐ Bachelor's Degree
☐ Advanced degree

Q. 25 What is your current employment status? Are you…

☐ Employed full-time
☐ Employed part time
☐ Self employed
☐ Currently unemployed
☐ Retired
☐ Homemaker

Q. 26 Including yourself, how many people live in your household?_____

IF MORE THAN ONE, ASK: How many of your household members are adults who are 18 years of age or older?_____

Q.27 Are you between …

☐ 18 to 29 years of age
☐ 30 to 39 years of age
☐ 40 to 49 years of age
☐ 50 to 64 years of age, OR
☐ 65 years of age or older

Q. 28 [DO NOT ASK: Record Gender: ☐ Male   ☐ Female]

That is all the questions I have today. Thank you very much for your time. If you have any questions about the study, please contact Margaret Petrella at 1-866-674-3971.

We will be calling you again about a week or two after you start using the technology to "check-in" and see how you are doing with the technology. We will also call you at the

end of the trial, in the first week or two of March to administer another survey. Are there any days of the week that are better for reaching you?
[RECORD DAY(S):_____]
Is there any particular time of day that is better for reaching you?
[RECORD TIME(S): _____]

Thanks again. Good bye.

RIAS EXTENDED USER EVALUATION
POST- SURVEY FOR BLIND AND VISUALLY IMPAIRED SUBJECTS

Hello, may I please speak with < first name / last name>? (*MUST speak to the participant*)

A. IF PARTICIPANT IS NOT AVAILABLE, IDENTIFY YOURSELF AND REQUEST A CALLBACK TIME:

Initial Call #	Date/Time	Callback Request #	Date/Time
Call 1	/	Callback 1	/
Call 2	/	Callback 2	/
Call 3	/	Callback 3	/
Call 4	/	Callback 4	/

B. IF PARTICPANT NEEDS TO COME TO THE PHONE:
Hi, am I speaking with < first name / last name>? (*Proceed with Introduction*)

IF PARTICIPANT ANSWERED PHONE:
INTRODUCTION
Hi, < first name>, my name is **(your first and last name)**. I'm calling on behalf of the U.S. Department of Transportation to ask you some questions about your use of transportation in the Seattle area. This is part of the study on Remote Infrared Audible Signage. Your participation in this interview is completely voluntary. Also, please remember that your responses are confidential, and your personal information, such as your name, will not appear in any of the reporting of the data.

Would this be a good time to ask you some questions?

[IF <YES>: Great. (Continue with interview below)]

[IF <NO, This is not a good time>: Can we schedule a more convenient time in the next few days?] <Record alternate date and time>_____

[IF <NO, I do not wish to participate>: Thank you for your time. We will no longer call you about this study. (Terminate call)]

Q. 1 First, our records indicate that you were given a RIAS receiver on [INSERT DATE: _____]. Does that sound right?

☐ Yes

☐ No → When did you obtain the receiver?_____

Q. 2 How often have you used RIAS since you obtained the receiver? [PROBE FOR DETAILS e.g. used on a regular basis, or just X times]

_____
_____
_____

Q. 3 Using a scale from 1 to 10, **where 1 = not at all satisfied** and **10 = extremely satisfied**, how satisfied were you with RIAS?

1	2	3	4	5	6	7	8	9	10	Don't Know	Refuse

Q. 4 For each of the following mobility areas, did your use of RIAS increase your level of confidence, decrease your level of confidence, or did it have no effect on your level of confidence? First...

	Increase	Decrease	No Effect	**DK**	**NA**
a. Using public transportation in general	1	2	3	8	9
b. Making transfers from one type of transportation to another type of transportation	1	2	3	8	9
c. Traveling in unfamiliar transit centers	1	2	3	8	9
d. Traveling in familiar transit centers	1	2	3	8	9

Q. 5a Did you use RIAS at King Street Station?

☐ Yes → ASK Q. 5b, 5c, 5d
Probe: Did you use RIAS as part of your typical travel at King St. Station, or did you go to King St. Station only to test RIAS? _____
☐ No → SKIP TO Q. 6a
☐ Don't know/Can't Recall [VOLUNTEERED] → SKIP TO Q. 6a

Q. 5b How helpful was RIAS in finding your bus or train at King Street Station?

☐ Very helpful
☐ Helpful
☐ Not too helpful
☐ Not at all helpful
☐ Don't know [VOLUNTEERED]

Q. 5c How helpful was RIAS in making transfers at King Street Station? [**RECORD RESPONSE THEN PROBE**: Can you describe the transfers you made at King St. Station?]

☐ Very helpful
☐ Helpful
☐ Not too helpful
☐ Not at all helpful
☐ Don't know [VOLUNTEERED]
☐ Does Not Make Transfers at King Street

Q. 5d Do you have any additional feedback on your experience using RIAS at King Street Station?
_____
_____
_____
_____

Q. 6a Did you use RIAS at Tacoma Dome Station?

☐ Yes → ASK Q. 6b, 6c, and 6d
Probe: Did you use RIAS as part of your typical travel at Tacoma Dome Station, or did you go to Tacoma Dome Station only to test RIAS?
_____
☐ No → SKIP TO Q. 7a
☐ Don't know/Can't Recall [VOLUNTEERED] → SKIP TO Q. 7a

Q. 6b How helpful was RIAS in finding your bus or train at Tacoma Dome Station?

☐ Very helpful
☐ Helpful
☐ Not too helpful
☐ Not at all helpful
☐ Don't know [VOLUNTEERED]

Q. 6c How helpful was RIAS in making transfers at Tacoma Dome Station [**RECORD RESPONSE THEN PROBE**: Can you describe the transfers you made at Tacoma Dome Station?]

☐ Very helpful
☐ Helpful
☐ Not too helpful
☐ Not at all helpful
☐ Don't know [VOLUNTEERED]
☐ Does Not Make Transfers at Tacoma Dome Station

Q. 6d Do you have any additional feedback on your experience using RIAS at Tacoma Dome Station?

_____
_____
_____
_____

Q. 7a Did you use RIAS at Everett Station?

☐ Yes → ASK Q. 7b, 7c, and 7d
   Probe: Did you use RIAS as part of your typical travel at Everett Station, or did you go to Everett Station only to test RIAS? _____
☐ No → SKIP TO Q. 8a
☐ Don't know/Can't Recall [VOLUNTEERED] → SKIP TO Q. 8a

Q. 7b How helpful was RIAS in finding your bus or train at Everett Station?

☐ Very helpful
☐ Helpful
☐ Not too helpful
☐ Not at all helpful
☐ Don't know [VOLUNTEERED]

Q. 7c How helpful was RIAS in making transfers at Everett Station [**RECORD RESPONSE THEN PROBE**: Can you describe the transfers you made at Everett Station?]

☐ Very helpful
☐ Helpful
☐ Not too helpful
☐ Not at all helpful
☐ Don't know [VOLUNTEERED]
☐ Does Not Make Transfers at Everett Station

Q. 7d Do you have any additional feedback on your experience using RIAS at Everett Station?

_____
_____
_____

Q. 8a Did you use RIAS at Bellevue Transit Center?

☐ Yes → ASK Q. 8b, 8c, and 8d
    Probe: Did you use RIAS as part of your typical travel at Bellevue Transit Center, or did you go to Bellevue Transit Center only to test RIAS? _____
☐ No → SKIP TO Q. 9a
☐ Don't know/Can't Recall [VOLUNTEERED] → SKIP TO Q. 9a

Q. 8b How helpful was RIAS in finding your bus at Bellevue Transit Center?

☐ Very helpful
☐ Helpful
☐ Not too helpful
☐ Not at all helpful
☐ Don't know [VOLUNTEERED]

Q. 8c How helpful was RIAS in making transfers at Bellevue Transit Center [**RECORD RESPONSE THEN PROBE**: Can you describe the transfers you made at Bellevue Transit Center?]

☐ Very helpful
☐ Helpful
☐ Not too helpful
☐ Not at all helpful
☐ Don't know [VOLUNTEERED]
☐ Does Not Make Transfers at Bellevue Transit Center

Q. 8d Do you have any additional feedback on your experience using RIAS at Bellevue Transit Center?

_____
_____
_____
_____

Q. 9a Did you use RIAS at Kent Station?

☐ Yes → ASK Q. 9b, 9c, and 9d
Probe: Did you use RIAS as part of your typical travel at Kent Station, or did you go to Kent Station only to test RIAS?
☐ No → SKIP TO Q. 10a
☐ Don't know/Can't Recall [VOLUNTEERED] → SKIP TO Q. 10a

Q. 9b How helpful was RIAS in finding your bus or train at Kent Station?

☐ Very helpful
☐ Helpful
☐ Not too helpful
☐ Not at all helpful
☐ Don't know [VOLUNTEERED]

Q. 9c How helpful was RIAS in making transfers at Kent Station [**RECORD RESPONSE THEN PROBE**: Can you describe the transfers you made at Kent Station?]

☐ Very helpful
☐ Helpful
☐ Not too helpful
☐ Not at all helpful
☐ Don't know [VOLUNTEERED]
☐ Does Not Make Transfers at Kent Station

Q. 9d Do you have any additional feedback on your experience using RIAS at Kent Station?

_____
_____
_____
_____

Q. 10a Did you use RIAS at Federal Way Transit Center?

> ☐ Yes → ASK Q. 10b, 10c, and 10d
> > **Probe:** Did you use RIAS as part of your typical travel at Federal Way Transit Center, or did you go to Federal Way Transit Center only to test RIAS?
>
> ☐ No → SKIP TO Q. 11
> ☐ Don't know/Can't Recall [VOLUNTEERED] → SKIP TO Q. 11

Q. 10b How helpful was RIAS in finding your bus at Federal Way Transit Center?

> ☐ Very helpful
> ☐ Helpful
> ☐ Not too helpful
> ☐ Not at all helpful
> ☐ Don't know [VOLUNTEERED]

Q. 10c How helpful was RIAS in making transfers at Federal Way Transit Center [**RECORD RESPONSE THEN PROBE**: Can you describe the transfers you made at Federal Way Transit Center?]

> ☐ Very helpful
> ☐ Helpful
> ☐ Not too helpful
> ☐ Not at all helpful
> ☐ Don't know [VOLUNTEERED]
> ☐ Does Not Make Transfers at Federal Way Transit Center

Q. 10d Do you have any additional feedback on your experience using RIAS at Federal Way Transit Center?

_____
_____
_____

Q. 11 If you have to buy your own RIAS receiver, which costs approximately $290, and the RIAS technology is available at the six Sound Transit stations where it is currently installed (including King Street, Bellevue Transit Center, Kent Station, Federal Way Transit Center, Tacoma Dome Station, and Everett Station), would you purchase the receiver?

> ☐ Yes
> ☐ No
> ☐ Don't know

Q. 11a Why or why not?

_____
_____
_____
_____

Q. 12 What about if the receivers were free, and the technology was installed at the same six transit stations? Would you use the technology?

    ☐   Yes
    ☐   No

Q. 13 What if the RIAS receiver cost $290 and RIAS were installed at **all** transit stations, platforms, and bus stops? Would you purchase the receiver?

    ☐   Yes
    ☐   No

Q. 13a Why or why not?_____
_____

Q. 14 Did your use of RIAS have any impact on:

	Yes	No	IF YES [please describe the impact]
a. Your ability to access different types of transportation?	1	0	
b. How likely you were to ask a transit operator or station agent for travel-related assistance?	1	0	
c. How likely you were to ask other passengers for travel-related assistance?	1	0	
d. Your social or recreational opportunities?	1	0	
e. The general quality of your life?	1	0	
f. Your trip-making behavior, such as the number or type of trips you made	1	0	

Q. 15 Did your use of RIAS provide you with any benefits? [Probe: Please describe:]

_____
_____
_____
_____

Q. 16 Based on your experiences using RIAS, do you have any issues or concerns about the technology? [PROBE: Did you have any problems using RIAS, either with the receiver or with the system more generally?]

_____
_____
_____
_____
_____

Q. 17. Based on your experiences using RIAS, do you have any recommendations for improvements to the technology or to the system?

_____
_____
_____
_____
_____
_____
_____

Q. 18 Do you have any other feedback on RIAS?

_____
_____
_____
_____
_____
_____
_____

That is all the questions I have today. Thank you very much for your time. If you have any questions about the study, please contact Margaret Petrella at 1-866-674-3971.

I have two quick housekeeping matters to ask you about.

1. Do you think you could return the RIAS receiver to the Sound Transit office on 401 South Jackson Street sometime in the next week or so?

    ☐ Yes

    ☐ No → I'll have Ella Campbell of Sound Transit contact you about arranging delivery of the receiver.

2. We plan on purchasing a Puget Pass for study participants. If we purchased a **May** Puget Pass, would that work for you? (e.g. you don't have an annual pass?) We can order the pass on April 15, and we will mail it to you as soon as we receive it. Could you please confirm your address:
   _____

Thanks again. Good bye.

DATE:
TIME:

Focus Group Discussion Guide: Blind/Visually-Impaired
(All Participants Not Formally Surveyed)

Hi, my name is Margaret Petrella. I work for the U.S. Department of Transportation.

Thank you for participating over the last several weeks in the study on Remote Infrared Audible Signage. We have brought you all together to hear about your experiences using Talking Signs.

Each of your experiences is important and valid, and they may or may not be similar to the experience of others.

I'll be asking you some questions about RIAS, but remember, there are no right or wrong answers – we just want to find out if this technology is useful.

Before we start,
let's take a few minutes to read the Study Description.

Participation in this extended discussion is completely voluntary, and if you do not wish to answer a question, you don't have to.

In order to protect your privacy, we will only ask for your first name, and we will not collect any other personal information.

I should let you know that J. Gayle Hays is here to take notes of our discussion, so it will be easier us to review what everyone says. We will not record your names, and will refer to each of you by a number.

When we report the data, no names will be used. We will simply summarize the overall findings from the discussion.

Finally, we are asking all of you to respect one another's privacy and NOT to discuss what you hear today outside this group.

Does anyone have any questions? Is everyone willing to participate?

- First, can we can go around the room and please tell everyone your **first** name and tell us about your use of public transportation – which stations you use and what types of trips you take using public transportation.

[NOTE: Each question and probe will be asked then each participant in round-robin fashion will be allowed to talk and respond to the question:]

- Can we go around the table again and everyone please give me a sense of how often or how many times you were able to use RIAS?

- Did your use of the technology have any affect on your trip-making behavior in general? PROBE: In what way?

  Did it affect your use of public transportation? In what way?

  > ADDITIONAL PROBES: Did it change the number of trips you made on public transportation? Did it change the places you traveled to?
  >
  > Did it change your use of paratransit (IF APPLICABLE)?
  >
  > Did it affect the type of trips you made on public transportation?

- Was the technology helpful to you?
    a. IF YES: In what ways was it helpful?
    b. IF NO: Can you explain why not?
    c. Did it assist or interfere with your existing wayfinding methods?

- **Does anyone here switch modes of transportation when they are making trips, for example from Sounder to the bus or from paratransit to the bus? [FOR PARTICIPANTS WHO MAKE MULTIMODAL TRIPS]:** You said that you sometimes have to switch from one type of transportation to another. Did this technology affect your ability to make those transfers? (PROBE: Did it make it easier or harder, or did it have no effect?)

- Overall, would you say your use of this technology had any impact on the general quality of your life?
    a. IF YES: Please describe. Can you give specific examples?
    b. Would you say this technology increases or decreases or has no effect on your ability to travel independently?

- Did you experience any problems using the technology? Please describe. (PROBE FOR SPECIFICS: Nature of problem, frequency etc.)

- Was the message set appropriate to your needs?

- Can you provide any recommendations or suggestions for ways to improve the technology? PROBE: Any others?

- If you have to buy your own RIAS receiver, which costs approximately $250.00 and the RIAS technology is available at six Sound Transit stations, would you purchase the receiver? Why or why not?
  (King Street, Bellevue Transit Center, Kent Station, Federal Way Transit Center, Tacoma Dome Station, Everett Station)

  PROBE: What about if the receivers were free, and the technology was installed at the same six transit stations?

  PROBE: What if the receiver cost 250.00 and RIAS were installed at **all** transit stations, platforms, and bus stops?

- Using a scale from 1 to 10, **where 1 = the worst and 10 = the best**, how satisfied were you with RIAS?

1	2	3	4	5	6	7	8	9	10	Don't Know	Refuse

- Do you have any additional feedback or comments that you would like to provide regarding your experience with this technology?

Thank you so much for your time. We really appreciate your willingness to participate in this research.

www.ingramcontent.com/pod-product-compliance
Lightning Source LLC
Chambersburg PA
CBHW081850170526
45167CB00007B/2964